Derrida on Time

'Perhaps there has been no more resilient form of thought in the last two hundred years than phenomenology, and yet, poignant critiques have been made of it. Joanna Hodge allows us to start to move beyond the horizon of phenomenology while recognizing the contributions it has made for thinking.'

Leonard Lawlor, *University of Memphis, USA*

Jacques Derrida was one of the most innovative and controversial thinkers of the twentieth century. In this outstanding book, Joanna Hodge explores the lasting significance of Derrida's legacy, and identifies as central to that legacy a rethinking of time.

Beginning with an exploration of the theme of time and temporality in Derrida's writings, Hodge discusses the transformation of philosophy resulting from Derrida's conception of time as non-linear, most notably expressed by the term *différance*. The author reads Derrida's writings as a response to those of Immanuel Kant and Edmund Husserl. Hodge addresses Derrida's radical disruption of Kant's determinations of time and explores Derrida's relationship to Husserl and the question of the destiny of phenomenology. By tracing Derrida's complicated departures from and returns to phenomenology, she also illuminates his relationship to such thinkers as Emmanuel Levinas and Maurice Blanchot, Paul Ricoeur and Jean Luc Nancy. As a result, Hodge argues, some of Husserl's concepts receive a decisive inflection, in Derrida's close attention to the registers and rhythms of enquiry, to Freudian analysis and to the singularity of literary register. Hodge then concludes by diagnosing in Derrida's writings a suspension of a decision between affirming a concept of history, and affirming a concept of time, emancipated from the burden of history.

Derrida on Time is essential reading for students of continental philosophy and literary theory.

Joanna Hodge is Professor of Philosophy at Manchester Metropolitan University and author of *Heidegger and Ethics* (Routledge 1995).

First published 2007
by Routledge
2 Park Square, Milton Park, Abingdon, Oxon OX14 4RN

Simultaneously published in the USA and Canada
by Routledge
270 Madison Ave, New York, NY 10016

Routledge is an imprint of the Taylor & Francis Group, an informa business

Typeset in Baskerville by Taylor & Francis Books
Printed and bound in Great Britain by
Biddles Ltd, King's Lynn

British Library Cataloguing in Publication Data
A catalogue record for this book is available from the British Library

Library of Congress Cataloging in Publication Data
A catalog record for this book has been requested

ISBN13: 978-0-415-43091-3 (hbk)
ISBN13: 978-0-203-94584-1 (ebk)

ISBN10: 0-415-43091-7 (hbk)
ISBN10: 0-203-94584-0 (ebk)

Derrida on Time

Joanna Hodge

Routledge
Taylor & Francis Group
LONDON AND NEW YORK

In memoriam
Jessie McDonald
(Aiken/Armstrong)
1889–1970

Contents

Preface

The various occasions on which I was fortunate enough to hear Jacques Derrida speak remain vivid in recollection. About ten years after the first of these, which was at the University of Essex in 1986, a presentation of what was to become *Of Spirit*, I formed the plan of writing about what I had learned from him. The enquiries which follow here are the outcome of this resolve. They started out from a formal delineation, tracing out a disruption and inversion of the order Kant attributes to his categories of quantity, quality, relation and modality, as capturing some truth about Derrida's enquiries with respect to time. With respect to quantity, Derrida condenses a certain number of concepts of time, those of Aristotle, Kant, Kierkegaard, Husserl, Heidegger and Freud, into a question of intensity: the excess of a psychic charge, over the capacities for registration of that charge. Originary impression thus exceeds expression. This parallels a move made by Jean-François Lyotard, emphatically, at the beginning of *Heidegger and the 'Jews'* (1986) but not only there, since the thought plays a role both in *The Differend: Phrases in Dispute* (1983) and in his *Lectures on the Analytic of the Sublime* (1991), which title masks the manner in which Lyotard proposes a reading of Kant, reversing the order of the three critiques. The implication is that the results of the Third Critique, on taste and judgment, disrupt and displace the theoretical positionings of the first two. An attempt might be made to assign priorities with respect to aspects of this, in the thinking of Lyotard and of Derrida. This would, however, distract attention from what is distinctive of each of these disruptive returns to a reading of Kant's critical philosophy, the one inflected more strongly through a response to Greek sophistry, and the other through a questioning of Kant's analyses of religion, but both informed by a response to Freud.

The reversal of the order between quantity and quality, and of the order of reading of Kant's critiques, generates an excess of meaning, which cannot be captured in an order of concepts. This repeats a finding Derrida makes in relation to Husserl's account of originary impressions and formulable meanings. Attention to this leads to a diagnosis of a delay in the formation of concepts and the movement of the distinctive displacement called *différance*. This displacement is then exaggerated in a second inversion of the ordering between relation and modality, whereby a modality, rethinking necessity in the 'perhaps', neither chance, nor determinism, takes precedence over relation, quality and quantity. This 'perhaps' as opening on to mischance erodes the distinctions between the categories and other concepts, so carefully set in place by Kant. Relation is marked as the chance encounter between text and response, and is subordinated to the timing of that chance. The 'perhaps' or 'happenstance' performed in the wanderings of Nietzsche's Zarathustra is then conjoined to the erratic movements, the 'free association' of Freudian analysis, and together they disrupt the classical order of possibility, actuality and necessity, as a hierarchy of increasing determinacy. The inversion of the order of Kant's exposition of the categories places a judgment of taste in advance of a transcendental aesthetic, and a hypothesis about self-forming matter in advance of a table of categories.

This inversion takes place suitably enough only in the margins of Derrida's *Truth in Painting* (1978), and by implication in the concurrent readings of Freud and of Nietzsche. In the former, a discussion of Hegel's *Aesthetics* is interrupted by a return to Kant's analyses of the beautiful and the sublime, and then by an implied encounter between the counter-posed analyses of art and politics proposed by Heidegger and by Benjamin. To excavate an account of this last encounter, between Heidegger and Benjamin, from Derrida's texts turned out to be beyond the reach of this set of enquiries, which instead return to the conjoining and disjoining of themes in the name of Kant. In place of a transcendental unity of apperception, Derrida offers a serial performance of self-constitution, or auto-thanato-biography. This can be traced out by paying attention to an order of reading and responding to the texts constitutive of a transmission of philosophy. This auto-crypto-thanato-biography has the form of a *circonfession* which takes the place of the order imposed in the name of transcendental unity, and of transcendental method more generally. Auto-bio-graphy as self-constitution thus occurs uniquely on each occasion of the processes involved in any registration of affectivity, of which reading is one mode. The reflexivity of the self, auto-, is in part hidden from itself, crypto-, and inscribes not only the life, bios-, but also the death, thanatos-, in writing, understood as a process interrupted but not terminated by death. *Différance* as the first term in the series of Derridean inventions takes the place of the delimitation of space and time in any transcendental aesthetics, and these processes of self constitution take the place of any critique of the judgment of taste. Other such (non)-terms would be writing, pharmakon, destinerrance, topolitology, duty, gift, subjectile, the *a-dieu*, hostipitality, *psyche*, auto-immunity, *literaterreur*, limitrophy. There is, however, no definite listing for these (non)-terms, and even less a definitive ordering for them.

A direct demonstration of this diagnosis unsurprisingly proved impossible to perform, and what follows has to be a *circonfession* of a *circonfession*. For the process of arriving at these schemata is not the same as a filling out of such schemata, and it is the former, not the latter which is given in what follows. What follows then is a response to the writings of Derrida, read as in the first instance a series of responses to Kant, which are blocked, deflected and rerouted by the chance encounters of those re-routings. This inevitably imposes a distortion on those writings, for which, no doubt, apologies are due to just about everyone, and first of all to Jacques Derrida, and to his companions in thought, Lyotard, Levinas, Heidegger, Husserl, Kierkegaard and Kant, and then to anyone who would have preferred the impossible book, to the studies it has proved possible to commit to paper. These studies follow a certain line of reading, of a certain line of reading. As a result a series of transformations of Kantian themes can be identified as indispensable to the thought in which they are carried along.

It is well known that the workings of an 'Idea in the Kantian sense' are identified by Husserl and emphasised in their readings of him by both Paul Ricoeur and Jacques Derrida. Conceptions of transcendental illusion and of unpresentability are taken up and transformed not just by Derrida but also by Jean-François Lyotard and by Jean-Luc Nancy. Accounts of experience, as held in place by distinct conceptions of limit conditions, set out the relations between a series of responses to and between the writings of Heidegger, Levinas and Blanchot. Furthermore, Kant's distinction between analysis of religion and theological dogmatism permits the differences between Walter Benjamin and Jacques Derrida to be placed firmly in the context of a questioning of an inheritance of the task of critique. Benjamin's inversion of Kant's proposal to think religion separately from theological commitments, thinking theological concepts independently of religious commitment, poses a different determination of the Kantian inheritance from that proposed by Emmanuel Levinas, who disruptively affirms the truth of Judaism, in place of Kant's affirmation of the truth of Christianity. Derrida, seeking to remain neutral between the religious and non-religious, and between Judaism, Christianity and Islam, can adopt neither of these moves. This permits him to unpick a series of occluded connections between negative theology and political theology, between religion and literature. In conclusion, a contrast opens out between a thinking of time, separated off from any conception of history, with Benjamin, and a thinking of history, separated off from conceptions of time, with de Man. As a result Benjamin perhaps surprisingly turns out to continue a line of enquiry first launched in the twentieth century by Husserl, of seeking to separate out an account of time, from naturalising presuppositions concerning history. The most startling of Husserl's innovations is that of an historical *apriori*, which appears to subvert both the historicality of history and the logical status of the *apriori*. Thus the historical *apriori* is a hybrid term, worthy of inclusion in the (non)-list of (non)-terms hazarded above.

The main claim to be examined here is that time is to be thought not as linear, but as curved, and that matter and its materiality are organised in accordance with asymmetrical relations arising from such curvature, rather than in accordance with a surmised line of continuous development from some notional beginning to some equally notional end point. Evidence for this claim is to be sought in a reconstruction of the relation of intentionality, which is basic to Husserl's phenomenology, as neither unidirectional, nor taking the form of a direct unswerving link between thinking and thought content. Instead intentionality is to be thought as a relation characterised by reversibility and swerve, as demonstrated in the movements characteristic of Derrida's readings of the texts constitutive of a certain transmission of philosophy: as marked by loopings, hesitations and precipitancy. The two relations basic to thinking time as linear, the before and after, and the 'now' of the transition from past to future and from future to past, are disrupted in advance by a syncopation of non-simultaneity, and by an after-shock which retrospectively determines the figuration of its sources.

Acknowledgements

In the course of the ten years during which I have attempted to formulate my questions, and to arrive at an acceptance of the nature of their impossibility, various happily timed invitations to speak, and to participate in conferences and symposia have provided a stimulus to further the struggle. I think especially of the International Philosophy Symposium in Alto Adige, in 1997, on *Specters of Marx*; the Irvine meeting of the IAPL, in 1998; and a conference on Derrida organised by graduate students at the University of Sussex in September 2003. I should also like to thank the various members of the Husserl Circle, and especially John Barnet Brough of Georgetown, Dermot Moran of UCD, and Nicolas de Warren of Wellesley College, for their hospitality, and for the scholarly rigour of its meetings, in Washington DC, in Dublin, and in Boston MA, in 2004, 2005 and 2006. The more orthodox readers of Husserl will not cease to be shocked by what Derrida and his admirers continue to do to the writings of Husserl. I have of course contracted other unpayable debts, not least to my colleagues in Manchester, who from time to time must have wondered if the story of a book about Derrida and Time could possibly have a happy ending: Paul Cammack, Martin Bell, Gary Banham, Ullrich Haase, Michael Garfield, Keith Crome, Mark Sinclair, Roxana Baiasu, Joaquim Siles Borras, doctoral and masters students; and indeed Wolfe Mays and David Melling, who bear some of the responsibility for appointing me to teach at MMU in the first place, and who, alas, are now dead.

I should like to acknowledge the very special importance of a reading group on Husserl, which met regularly from 2001 to 2003 in the office of Alex Samely at the University of Manchester, without which and without whom the reading of Husserl proffered here would be so much the poorer. I also owe much to a conversation about how to read Derrida, between myself and Sean Gaston, which started in autumn 2003, and, again, without which even more would have remained shrouded in obscurity. I should also like to thank a number of anonymous readers of proposals and drafts who have no doubt attempted to improve matters. The librarians in Manchester, at the All Saints Library and the John Rylands, and those in the Bodleian, in Oxford, and especially those at the Taylor, have proved over and over again how kindness and courtesy help in the face of even the most obdurate impasse.

I should also like to thank the AHRC, for support in 2003–4, to whom the debt finally fell due in December 2005.

Joanna Hodge
December 2006

Abbreviations

B&D	Geoffrey Bennington and Jacques Derrida, *Jacques Derrida*.
B: F	Maurice Blanchot, *Friendship*.
B: FF	Richard Cohen (ed.) *Face to Face with Levinas*.
B: SB	Maurice Blanchot, *The Step (not) Beyond*.
B: WD	Maurice Blanchot, *Writing of the Disaster*.
B: WF	Maurice Blanchot, *The Work of Fire*.
Benjamin SW	Walter Benjamin, *Selected Writings* vols 1–4.
CJ, AK V	Immanuel Kant, *Critique of Judgment*.
Confessions	Augustine, *Confessions*.
D: 1997, AA	Jacques Derrida, 'The Animal Which Therefore I Am (More to Follow)', *Critical Inquiry*, vol. 28, no. 2, winter 2002.
D: 1986, AC	Jacques Derrida, 'Aphorism Countertime', in Jacques Derrida, *Acts of Literature*, ed. and trans. Derek Attridge.
D: 1996, AD	Jacques Derrida, *Adieu to Emmanuel Levinas*.
D: 1993, AP	Jacques Derrida, *Aporias: Dying – Awaiting (One Another at) the 'Limits of Truth'*.
D: 2006, AQD	Jacques Derrida, *L'animal que donc je suis*.
D: 2002, AR	Jacques Derrida, *Acts of Religion*, Gil Anidjar (ed.).
D: 2003, B	Jacques Derrida, 'Maurice Blanchot est mort', in Christophe Bident and Pierre Vilar (eds) *Maurice Blanchot: récits critiques*.
D: 1991, BTB	*A Derrida Reader: Between the Blinds*, Peggy Kamuf (ed.).
D: 1992, DNT	Harold Coward and Toby Foshay (eds) *Derrida and Negative Theology*.
D: 2000, EC	Jacques Derrida, 'And etc.', in Nicholas Royle (ed.) *Deconstruction: A Critical Reader*.
D: 2004, EU	Jacques Derrida, *Eyes of the University: Right to Philosophy 2*.
D: 1983, G1, BTB	Jacques Derrida, 'Sexual Difference, Ontological Difference; *Geschlecht* One', in Kamuf (ed.) *A Derrida Reader: Between the Blinds*.
D: 1985, G2	Jacques Derrida, '*Geschlecht* II: Heidegger's Hand', in Sallis (ed.) *Deconstruction and Philosophy*.

D: 1989, G4	Jacques Derrida, 'Heidegger's Ear: Philopolemology: Geschlecht Four', in Sallis (ed.) *Reading Heidegger: Commemorations*.
D: 1992, GD	Jacques Derrida, *The Gift of Death*, trans. David Wills.
D: 1991, GT	Jacques Derrida, *Given Time 1: Counterfeit Money*.
D: 1989, HA	Jacques Derrida, 'How to Avoid Speaking: Denials', in Coward and Foshay (eds) *Derrida and Negative Theology*.
D: 1962, HOG	Jacques Derrida, *Introduction to Husserl's 'Origin of Geometry'*.
D: 1982, MC	Jacques Derrida, 'Mes Chances: On Some Epicurean Stereophonies', in Smith and Kerrigan (eds) *Taking Chances*.
D: 1972, MP	Jacques Derrida, *Margins of Philosophy*.
D: 1967, OG	Jacques Derrida, *Of Grammatology*.
D: 1991, OH	Jacques Derrida, *The Other Heading*.
D: 1987, OS	Jacques Derrida, *Of Spirit: Heidegger and the Question*.
D: 2000, OT	Jacques Derrida, *On Touching: Jean-Luc Nancy*.
D: 1980, PC	Jacques Derrida, *The Post Card*.
D: 1994, PF	Jacques Derrida, *Politics of Friendship*.
D: 1954, 1990, PG	Jacques Derrida, *The Problem of Genesis in Husserl's Philosophy*.
D: 2001, PM	Jacques Derrida, *Paper Machine*, trans. Rachel Bowlby.
D: 1989, *Psyche*	Jacques Derrida, *Psyche: Inventions of the Other*, in Lindsay Waters and Wlad Godzich (eds) *Reading De Man Reading*.
D: 2003, RS	Jacques Derrida, *Rogue States: Two Essays on Reason*.
D: 1993, SM	Jacques Derrida, *Specters of Marx*.
D: 1967, 1968, SP	Jacques Derrida, *Speech and Phenomena and Other Essays*.
D: 1995, T	Jacques Derrida, 'Tense', in Kenneth Maly (ed.) *The Path of Archaic Thinking: Unfolding the Work of John Sallis*.
D: 1978, TP	Jacques Derrida, *Truth in Painting*.
D: 2001, WA	Jacques Derrida, *Without Alibi*.
D: 1967, WD	Jacques Derrida, *Writing and Difference*.
De Anima	Aristotle, *De Anima*.
GA 1–99	Martin Heidegger, *Gesamtausgabe* (Frankfurt am Main: Vittorio Klostermann 1976–).
H: BPP	Martin Heidegger, *Basic Problems of Phenomenology* (GA 24).
H: FCP	Martin Heidegger, *The Fundamental Concepts of Phenomenology: World, Finitude, Solitude* (GA 29–30).
H: MFL	Martin Heidegger, *Metaphysical Foundations of Logic* (GA 26).
H: OBT	Martin Heidegger, *Off the Beaten Track* (GA 9).
H: SZ/MR	Martin Heidegger, *Being and Time*.
HUA	Husserliana; Edmund Husserl, *Werke*, The Hague: Martinus Nijhoff, Dordrecht: Kluwer Academic Publishers and Springer Verlag.
HUE	Husserl's Collected Works in English vols 1–12, Dordrecht: Kluwer Academic Publishers and Springer Verlag.

Ideas One	Edmund Husserl, *Ideas: A General Introduction to Pure Phenomenology.*
Ideas Two	Edmund Husserl, *Ideas Pertaining to a Pure Phenomenology,* second book.
Kant: A/B	Immanuel Kant, *The Critique of Pure Reason.*
L: DE	Emmanuel Levinas, *Discovering Existence with Husserl.*
L: EE	Emmanuel Levinas, *Existence and Existents.*
L:GCM	Emmanuel Levinas, *Of God Who Comes to Mind.*
L: GDT	Emmanuel Levinas, *God, Death and Time.*
L: OB	Emmanuel Levinas, *Otherwise than Being, or, Beyond Essence.*
L: LR	Sean Hand (ed.) *The Levinas Reader.*
L: RL	Robert Bernasconi and Simon Critchley (eds) *Re-reading Levinas.*
L: TI	Emmanuel Levinas, *Totality and Infinity: An Essay on Exteriority.*
Ly: D	Jean-François Lyotard, *The Differend: Phrases in Dispute.*
Metaphysics	Aristotle, *Metaphysics.*
Physics	Aristotle, *Physics.*
PFL	Penguin Freud Library.
R: HP	Paul Ricoeur, *Husserl: An Analysis of His Phenomenology.*
R: TN	Paul Ricoeur, *Time and Narrative.*

In the beginning

1.1 On time and movement

In the beginning:

> After this discussion, the next topic to look into is time. It makes sense to start by rehearsing the difficulties which the issue generates, and in doing so we will draw on non-specialist ideas as well. The question is, first, whether or not it is a real entity, and second, what its nature is.
>
> (*Physics* iv.10. (217b 29))

If time is the measure of movement (Aristotle, *Physics* 219b 36), and if movement takes a number of mutually irreducible forms, the measure of movement will fragment and time too will take a number of irreducible forms.[1] The enquiries which follow explore various aspects of the paradoxes generated by such a hypothesis, not least the disruption of linear history and of a history of philosophy which divides moderns from ancients. Ever since Heidegger's early analyses, before the move to Marburg, there has been available the strong interpretation of Husserl's rethinking of time, as a retrieval of Aristotle's analyses of time, and as an account of the movements of the soul, whereby concepts of both soul and movement are transformed.[2] Derrida's writings contribute to charting the outlines of this transformation, and of the emergence of a definitive break between ancients and moderns. Heidegger subsequently seeks to move the point of contact with the Greeks back from Aristotle to Parmenides and Anaximander, to a Greek philosophy less dominated by a Latin language reception. For Derrida, however, the connection must go by way of St Augustine, and his conversion from paganism, and by way of the writings of Sigmund Freud, and his reconversion to paganism. The recasting of an account of the soul in the thoroughly secularised terms of Freud's topographies has a neglected importance here.[3] Derrida's Husserl is to be found between an Augustine emergent from paganism and a Freud emergent from 3,000 years of theocracy.

The movements of thought, as *psyche*, are invoked in the *Physics*, at 218b 27, but Heidegger attends more to the Aristotle of *De Anima*.[4] Derrida returns to this connection between a movement of time and the movements of soul, or spirit, on a number of occasions, in his pursuit of a genealogy for the Freudian concept of *psyche*.[5] This concept and its genealogy set out a distance between a more open conception of consciousness, and one subordinating consciousness, necessarily, to a conception of subjectivity. Thus the line of discontinuity and swerve, from Aristotle, through Augustine, to Husserl and Freud may be thought to be opposed to a continuous line, through Thomas, Descartes and Hegel, leading into the dogmatics of concepts of subjectivity. However, when Derrida discusses Aristotle on time, in the essay, '*Ousia* and *Gramme*, Substance and Line, Note on a Note in *Being and Time*' (1968), he makes the connection not to Husserl, but to Hegel, and to Heidegger's attempted rewriting of Husserl, on time and being, thus apparently consigning Husserl to history.[6] This deletion of Husserl in favour of Hegel and of Heidegger requires some explanation, since it can be shown that nevertheless throughout Derrida's enquiries, especially but not only when he reads Freud and Blanchot, Levinas and Heidegger, Marion and Nancy, the central figure and thematics are that of Husserl and of Husserlian phenomenology. The claim to be made out here is that the resources for analysis provided by Husserl, by contrast to those provided by Heidegger, and by Hegel, assist Derrida in resisting the tendency within philosophy and phenomenology to reinstall a privilege to Christianity as a religion, and to Christian onto-theological preoccupations. This comes into focus in Part IV of these enquiries, in which a splitting of religion from its circumscription within theology offers an opportunity to undo the spontaneous preferral of one religion over others.

Part II, after this introduction, considers Derrida's readings of Husserl, and Part III will consider the writings of Levinas and of Heidegger, as responses to Husserl, setting out Derrida's responses to each as further developments of his readings of Husserl. Part IV, as stated, will consider the relation between religion and theology, and Part V delineates the implications of Derrida's invocation of an animality, inherited by descent, and its embedding in the machinic structures of the inorganic, from which this animality emerges, and to which it returns: ashes to ashes. It will make a connection from this back to the notion of *circonfession,* emergent in the juxtaposition of Derrida's responses to Augustine and to Freud, which takes up and reinscribes Husserl's account of time and his notion of the living present. Husserl's account of inner time consciousness was disputed already by Heidegger, in his lectures of summer 1928, *The Metaphysical Foundations of Logic* (GA 26: 1928), where he writes: 'That which Husserl still calls time-consciousness, i.e. consciousness of time, is precisely time itself, in the primordial sense' (H: MFL p. 204).[7] In these lectures, Heidegger announces the publication of a version of Husserl's first account of time, *The Phenomenology of Internal Time-consciousness* (1928), and, as their editor, Heidegger was in a good position to know the difficulties preventing Husserl from completing that account.

As is well known, Levinas worked on the phenomenologies of both Heidegger and Husserl; also well known is Heidegger's difficult relationship to Husserl. Both Levinas and Heidegger depart from Husserl's phenomenology; they also both set out critiques of the philosophy of Hegel. For while in the 1920s, Heidegger discusses Husserl's phenomenology, as a response to the enquiries opened up in *De Anima*, concerning the movements of thought, in the 1930s he addresses himself also to Hegel's response to this text, when he turns to a reading of the *Phenomenology of Spirit* and to a more lengthy engagement with German idealism.[8] This delay in engaging with Hegel may be the consequence of Hegel's presumption, not shared by Husserl and Heidegger, that the invention of a conception of subjectivity is unequivocally an advance for thinking. Certainly, Hegel's affirmation of a Christianised version of Aristotle, proposed by Thomas in the Middle Ages, is also in dispute, for already in the 1920s, Heidegger seeks to distance philosophy from theology.[9] Hegel, however, engages more directly than Husserl in questions of ontology, to which, in Heidegger's view, Husserl's phenomenological enquiries fail to provide adequate access.

Derrida's responses to Husserl are inflected by his readings of both Heidegger and Levinas, and indeed by his reading of a French version of Hegel, in which the notion of dialectics is construed as open, and as still in process, as opposed to one closed off within an actualised absolute. Hegel concludes his *Philosophy of Spirit*, volume 3 of the *Encyclopaedia of the Philosophical Sciences*, by quoting from book *lamda* of Aristotle's *Metaphysics*, concerning the object of thought, as that which is intrinsically best, leading up to the definition of the properties of the divine as life, continuity, and eternal duration (*Metaphysics* 1072b 33).[10] Intriguingly, in the immediately following chapter, there is an argument for there being forty-seven, or possibly forty-nine prime movers, depending on an enumeration of autonomous, spherical movements, and thus forty-seven, or possibly forty-nine distinct movements to be measured. This, of course, accords with hypotheses concerning a multiplication of dimensions of space, beyond the standard three, with all the disruptive effects that has on notions of time as single, continuous, orderly succession. In this context, Hegel's decision to identify just one prime mover begins to look arbitrary. This sets up a division within the history of philosophy, between reading that tradition as a reception of Aristotle's inaugurating onto-theology, proposed in these definitions of the properties of the divine, as Christianised by Thomas, and as systematised by Hegel, versus one provided by Husserl, in which philosophical status is secured to figures in the past, in so far as they anticipate the discoveries of phenomenology. It will be important in what follows to distinguish between Hegel's subordination of phenomenology to onto-theology, and the converse move made by both Husserl, and, appearances to contrary, Heidegger, who make ontological commitment the consequence of the discoveries of phenomenological analysis.

The writings to be discussed in the following enquiries are in the main taken up with exploring the diminishing prospects of completing phenomenological enquiry, such that specifications of ontological commitment might be thought to arise from it. For some, this is a mark of failure, but I shall suggest that for Derrida, and especially with respect to an understanding of time, the inability to pursue phenomenological analysis to such a conclusion permits the consequentially emergent, more sophisticated thinking of time to be recognised for what it is. The paradoxes arising from attempts to thematise time are the occasion for this series of enquiries, and the inability of either Hegelian ontology or the phenomenologies of Husserl and Heidegger to render time adequately as a theme will be taken to be the occasion for philosophical celebration, not a source of regret. These enquiries proceed by paying attention to the manner in which modes of reading and discussing texts set up a series of distinct movements in which time is differentially articulated, in Part II in relation to Derrida's readings of Husserl, with interruptions from Freud; in Part III, in relation to reading of Heidegger, Levinas and Blanchot; in Part IV in relation to a reading of Benjamin, and to a reading of Derrida, reading Benjamin; and in Part V in relation to the *animot* of *L'animal que donc je suis* (2006).[11]

The possibility of there being radically distinct forms of movement, and of there being irreducibly distinct forms of measurement for them, is to be pursued in the first instance by close attention to the manner in which Derrida reads the writings of Husserl, but it may also be traced out in the movements constitutive of those analyses themselves, and in the various responses to them, in which the phenomenological tradition of the twentieth century consists. Thus the readings of Husserl offered by Levinas and Heidegger, by Marion and Nancy develop and display a variety of movements opening out in Husserl's enquiries, not all of which are compatible with one another. The suggestion is that the task of ontological constitution, in this case of the nature of time, follows the profile of the textual proliferation, prompted by Husserl's enquiries. That textual proliferation, in turn, grounds in the eidetic structures of time, of the distinct temporalities of various distinct domains of entities, and of temporalisation, which may be thought to be given *apriori*, but which, perhaps, are not fully retrievable, within the texts and thoughts of finite human articulations. These *apriori* givens may be thought to haunt those texts and thoughts as traces of an inaccessible, humanly impossible ontological fulfilment. With both Levinas and Marion, the mode in which to think a disruption, in phenomenology, of a finitude of human thinking by the arrival of non-finite thoughts of infinity and eternity, is in dispute, and this is, in turn, disputed both by Heidegger, and by Jean-Luc Nancy, for whom time arrives as finite, or, in some sense, finished.[12] For concepts of infinity are not themselves infinite. In this dispute, Jean-Luc Nancy and Jacques Derrida find themselves aligned with Heidegger and against Levinas, and Marion, and, indeed, Walter Benjamin.

In the first part of these enquiries, the focus for attention is Derrida's readings of Husserl, both those from 1953–67, and the return to a concern with Husserl, from the time leading into the publication in 1990 of the 1953–54 thesis, *The Problem of Genesis in Husserl's Philosophy* (1990), culminating with the publication of *On Touching: Jean-Luc Nancy* (2000).[13] The earlier readings prepare the way for the formation of the terms *écriture*, or writing; *trace*, borrowed from Levinas' reading of Husserl; and *différance*, developed out of Heidegger's reading of Husserl, the latter in *Of Grammatology* still called an 'economic concept'.[14] The term through which these three are held together is that self-subverting notion of grammatology, the study of the line (*gramme*), or of letters (*grammata*), which takes place in advance of the line, or letter. For the *logos* of the line or letter is to be framed by the chiasmatic invocation of the time and tense of the line or letter. In the preface to *Of Grammatology*, there is to be found the following remark:

> Although the word 'age' or 'epoch' can be given more than these determinations, I should mention that I have concerned myself with a *structural figure* as much as a *historical totality*. I have attempted to relate these two seemingly necessary approaches, thus repeating the question of the text, its historical status, its proper time and space. The age already in the *past* is in fact constituted in every respect as a *text*, in a sense of these words that I shall have to establish. As such the age conserves the values of legibility and the efficacy of the model and thus disturbs the time of the line or the line of time. I have tried to suggest this by calling upon and questioning the declared Rousseauism of a modern anthropologist.
>
> (D: 1967, OG p. xc)

The hypothesis of a disjunction between the forms of structural figure and of historical totality will require some attention, as will the implied transition between future and past, which 'is in fact constituted in every respect as text'. The future, then, by implication, is that which will have been written, or will have been 'in fact constituted in every respect as text'. Historical or textual totalities are disrupted in their claims to completeness by an irreducible naming function, 'Husserl', for example, or 'Heidegger' or even, paradoxically, 'Kierkegaard'. These names pick out structures of continuous but not necessarily internally consistent structures of enquiry. These partial structures are not to be contained in some further systematic totalising context, but are rather open to a non-finite series of re-deployments and re-reading.

Derrida's readings of Levinas' disjunctive thinking of totality and infinity disrupts the presumption that structure and history forms a basic opposition, and displaces the opposition between future and past, in favour of a reflection on differences between finite and non-finite time. The determinations of time in tense structure and in the temporal sequence, past, present, future, are to be contrasted with a thinking of the possibility of determining time at all, for which Husserl introduces the notions of temporality, and temporalisation, as means for describing the conditions for the appearing of time, in any such determinate mode. I propose to take this remark from *Of Grammatology* as a guide to reading Derrida's writings, pausing only to substitute the phrase 'an undeclared Aristotelianism of a certain phenomenology', for that concerning the declared Rousseauism of Lévi-Strauss. This stretches the framing of context from that of a post-Cartesian meditation, with Husserlian phenomenology taken to be a response to and critique of Descartes, to one which takes Husserl to respond to Aristotle's surveys of philosophical thinking among the Greeks, for which philosophy has always already begun. Husserl, in *The Crisis of the European Sciences and Transcendental Phenomenology* (1936), makes it clear that he supposes there to have been an event of thinking among the Greeks which sets out an inheritable practice of infinitisation, with respect to thought, and a practice of formalisation, with respect to analysis.[15] He and Heidegger agree that there is a danger of losing, eliding or forgetting some originary insight made available at this time, and they jointly suppose the task of philosophy to be an overcoming of this tendency to lose contact with these originating insights. Husserl's distinctive contribution here is the refinement of a notion of idealisation, as the bearer of meaning across contexts of enquiry, and of formalisation, as distinct from generalisation.[16] This attention to the history of the discipline contrasts both to Kant's notion of a critical delimitation for concepts, once and for all time, and to Hegel's interiorisation of a passage of time within the movements of the concept as spirit.

When Derrida in *Of Grammatology* entitles a section 'Writing before the letter', he marks up the paradox that any conceptual formation to be analysed, in any theoretical or philosophical undertaking, has the form it will be revealed to have in advance of its exposure to view. Ontology predates phenomenology, since the delimitability of concepts pre-dates their explicit formulation, and implicitly regulates a domain of enquiry in which attempts can then be made to render them thematic and determinate. This non-simultaneity of conceptual form and conceptual formulation is a function of a paradoxical time of conceptual articulation, which comes to the fore already in the writings of Husserl, and perhaps most explicitly as the notion of pre-predicative experience, in the text, jointly authored with Ludwig Landgrebe, *Experience and Judgment: A Genealogy of Logic* (1938), which, according to Derrida, was first conceived and sketched in 1919.[17] This non-simultaneity, and delayed arrival, of conceptual lucidity from a pre-given, but undisclosable source is distinctive of the temporalities of enquiry within which human beings function. This structure prompts Derrida to make the move, outraging Husserl scholars, of introducing Freud into his discussion of Husserl's analyses of time and the other, in *Speech and Phenomenon: An Introduction to the Problem of Signs in Husserl's Phenomenology* (1967).[18] In this text, Derrida deploys the Freudian concept of the delayed affect, the *après coup* of *Nachträglichkeit*, literally, the mode of the after-impact, which marks the arrival in consciousness of a recognition of an affect, or concept, previously imprinted in some register other than that of conscious awareness.[19] This marks the irreducible knotting together for Derrida of a time before the time of recognition, and a time of affective articulation prior to the time of rational delimitations.

At this point there must be marked as an open question, whether the Husserlian notions of consciousness are to be restricted to those of conscious awareness, or whether, with Freud and with Aristotle, there are levels of psychic functioning, for Husserl, contributing to the constitution of consciousness, of which consciousness need not, or, indeed, cannot be aware. This opens out the gap between conceptions of consciousness, and its possible self-maskings, on the one hand, and, on the other, those of subjectivity which can be made transparent to itself. Husserl begins to open out these distinctions in the analyses of time, dating back to the Göttingen years, in the 1904–5 lectures on time, and develops them subsequently in the distinctions between static and genetic constitution, in the drafting of the two lines of enquiry for his *Ideas: A General Introduction to Pure Phenomenology* (1913, 1952).[20] This then is further developed in the distinction between passive and active synthesis, brought to the fore in the 1920–21 lectures: *Analyses of Passive and Active Synthesis: Introduction to Transcendental Logic*

(HUE 9, 2001).[21] The Freudian and Aristotelian notions of the *psyche* disrupt the self-presence of a Cartesian self-introspecting subjective consciousness, and they also disrupt the attempt in the phenomenology of Merleau-Ponty to privilege the time of natural consciousness and its perceptions, over that of ideal consciousness and its images. This disruption is pursued by Derrida in all the writings collected under the title *Psyche: inventions de l'autre* (1987–).[22] Derrida's reading of Husserl, in 1967 is, however, already interrupted by the arrival of this extraneous name: Freud.[23]

The juxtaposition of the names, Husserl and Freud, reveals a certain complicity with respect to a retrieval into the present of contents, affects, impressions or originary meanings, for which there is no definite previous time of registration. The essay 'To speculate – on "Freud"', in *The Post Card: From Socrates to Freud and Beyond* (1980), demonstrates a disruption of Freud's own reading of the pleasure principle, and of pleasure itself, by the arrival of trauma and the death drive.[24] Texts may then already be disrupted from within, with the extraneous name merely marking that internal hiatus. The structure of *The Post Card* intimates this, since the first half is made up a series of autobiographical diary entries, to which a series of loosely connected essays, concerned with aspects of Freud's psychoanalytical writings, are then appended. In the first part of that text, '*Envois*', Derrida introduces two notions of time, important for this reading of his writings: singular anachrony and what he calls *restance*, the remainder which resists thematisation. He does this by conjoining the names Freud and Heidegger:

> Here Freud and Heidegger, I conjoin them within me like the two great ghosts of the 'great epoch'. The two surviving grandfathers. They do not know each other, but according to me they form a couple, and in fact just because of that, this singular anachrony. They are bound to each other without reading each other and without corresponding.
>
> (D: 1980, PC p. 191)

The time of a singular anachrony is that of a determinate non-simultaneity, where the two temporal determinations are linked, inseparable, and marked by a rhythm of syncopation, which disjoins any attempt to grasp them in a single articulation. In relation to this conjunction of names, Derrida articulates the resulting temporal schema in the following way:

> The master thinkers are also masters of the post. Knowing well how to play with the *poste restante*. Knowing how not to be there and how to be strong for not being there right away. Knowing how not to deliver on command, how to wait and to make wait, for as long as what there is, that is strongest within one, demands – and to the point of dying without mastering anything of the final destination. The post is always *en reste*, and always *restante*. It awaits the addressee who might always, *by chance*, not arrive.
>
> (D: 1980, PC p. 191)

This non-arrival is the key to his differences with Lacan, who reads Freud on the unconscious, to the effect that the unconscious affect must always arrive at its destination.[25]

The notion of a '*restance*' returns in the essay on Freud, in relation to the analysis of the death drive, as unpresentable:

> Such would be the de-monstration. Let us not abuse this facile play on words. The de-monstration makes its proof without showing [*montrer*], without offering any conclusion as evidence, without giving anything to carry away, without any available thesis. It proves according to another mode, but by marching to its *pas de demonstration*. It transforms, it transforms itself in its process rather than advancing the signifiable object of a discourse. It tends to fold into itself everything that it makes explicit, to bend it all to itself. The *pas de demonstration* is of that which remains in this *restance*.
>
> (D: 1980, PC p. 296)

The *pas de demonstration* displays a link to the double meaning of the 'pas' of the step which is a negation, as analysed by Maurice Blanchot in *Le pas au delà* (1973) translated as *The Step (Not) Beyond* (1992).[26] This in turn connects to the analyses of negative theology which mark the trajectory of Derrida's published writings through the 1980s into the 1990s, to be discussed in the fourth part of these enquiries. Derrida marks up a difference between a logic of linear argument, in accordance with deduction, induction, and analysis of wholes into parts, and a logic of the symptom, in which no such linear order can be expected, and in which a series of *mises en abîme* prevents the ordering of parts into a single structural order. This singular anachrony intimates that the thinking of each, of Freud and of Heidegger, is characterised by a certain openness to disruption, and by an inconclusiveness, with earlier and later theorising not entirely congruent with one another, left disruptively standing alongside one another.

Derrida's reading of Husserl, in the 1953–54 thesis, is disrupted by the pursuit of and deployment of Husserl's own notion of 'the Idea in the Kantian sense', in order to demarcate four distinct phases of the reading of Husserl. These two disruptions of Husserl's texts, by Husserl's own invocation of Kant, and by Derrida's importation of the name, Freud, generate an elliptical movement into Derrida's reading, which subsequently entails that whenever he later introduces considerations of Freud, or indeed of Kant, there is an implicit reference back to this earlier encounter, drawn out over fifteen years, with Husserl. In the first instance, this elliptical movement rotates the reading of Husserl around the two names, Kant and Freud, to reveal an unclarified origin in pre-predicative experience, and an unrealised end point, the fulfilment of an intuition of the Idea in the Kantian sense. This elliptical movement then takes on that other feature of ellipses, of eliding reference to indispensable parts of a single structure, or of an argument. This semantic feature of ellipsis is taken up by Jean-Luc Nancy, in his discussion of the writings of Derrida, from 1987, 'Elliptical Sense': '(One can easily imagine how "to speak on ellipsis" might be a statement of a concept of Derrida himself)'.[27] This elliptical movement, if not this elliptical sense, provides access to what is distinctive about Husserl's analyses of time, and of temporalisation, as the movement through which time arrives as object of thought.

The relation of this temporalisation, as the movement of thinking itself, to the distinct temporalities of time, as objects of thought, also approximates to that of an ellipsis, as does the task of describing this relation. This is sketched out by Husserl in the later accounts of time in terms of a necessary correlation between *noeses* as processes of thought, and *noemata* as their objects. This is the rewriting, in *Ideas One* (1913), of the relation introduced in *Logical Investigations* (1900/01) under the rubric 'intentionality'. Derrida himself writes of this elliptical movement, in his 1967 essay, 'Form and Meaning: A Note on the Phenomenology of Language':

> Thus one does not have to choose between two lines of thought. Rather, one has to meditate upon the circularity which makes them pass into one another indefinitely. And also by rigorously repeating this *circle* in its proper historical possibility, perhaps to let some *elliptical* displacement be produced in the difference of repetition, a deficient displacement doubtless, but deficient in a way that is not yet – or is no longer – absence, negativity, non-Being, lack, silence. Neither matter nor form, nothing that could be recast by some *philosopheme*, that is, by some dialectics, in whatever sense dialectics may be determined. An ellipsis both of meaning and of form: neither full speech, nor a perfect circle. More a less, neither more nor less. Perhaps an entirely other question.
>
> (D: 1972, MP p. 173)[28]

Derrida thus rewrites the linear unidirectional relation of intentionality between thinking, and its objects as an elliptical movement, which neither fully reaches an object, nor fully retrieves its pre-predicative conditions, in thinking. This leads into a disruption of Husserl's analyses of time. The ellipsis with which Derrida is principally concerned in this paper is the elision of the question of what is, in the Husserlian insistence on attention to formalisation. However, he also here develops this semantic ellipsis into a diagnosis of a movement distinctive of Husserl's thought. Derrida's readings of Husserl also take this elliptical movement, and attention to the form of Derrida's analyses provides guidance to what he supposes the contents of Husserl's analyses to be. For the form of the analyses is inseparable both from the problems they seek to analyse, and from the problems they pose, in a tension between the commitment to *apriorism*, while also sustaining a commitment to tracing out genesis; and in the tension between the workings of an empirical consciousness, aware of itself, and a transcendental consciousness, which, as absolute, does not permit of the regress. This last is provided with layers of articulation, above all else of inter-subjective and perhaps inter-generational connectedness, which cannot thus be rendered transparent to its supposed bearer, as individuated empirical consciousness.

In a paper from 1964, first delivered in 1959, '"Genesis and structure" and Phenomenology', Derrida draws attention to Husserl's preoccupation with rewriting Kant's conception of transcendental aesthetics.[29] In the midst of a summary of the results of his reading of Husserl so far, Derrida writes:

> Thus at these two poles of opening and from within the very transcendental structure of all consciousness there would arise the necessity for the transition to a genetic constitution and for the new 'transcendental aesthetic' which will be announced unceasingly but will be deferred always, and within which the themes of the Other and of Time were to have permitted their irreducible complicity to appear. It is that the constitution of the other and of time refers phenomenology to a zone in which its 'principle of all principles' (as we see it, its metaphysical principle: the *original self-evidence* and *presence* of the thing itself in person) is radically put in question. In any event, as can be seen, the necessity of this transition from the structural to the genetic is nothing less than the necessity of a break or a conversion.
>
> (D: 1967, WD pp. 163–64)

This 'principle of all principles', announced by Husserl in section 24 of *Ideas One*, is open to a number of different interpretations, but it affirms a form of direct access from intuition as a 'pure seeing', to a self-evidence of essences, which constrains the possible presentations of phenomena.[30] Shortly before this, Derrida has invoked an irreducible difference between wisdom and knowledge:

> This irreducible difference is due to an interminable delaying (*différance*) of the theoretical foundation. The exigencies of life demand that a practical response be organized on the field of historical existence, and that this response precede an absolute science whose conclusions it cannot await.
>
> (D: 1967, WD p. 161)

The shift from the strictly Husserlian contrast between static and genetic phenomenology, to that between structural and genetic considerations requires some commentary, but in outline the move made here is clear enough. Husserl's recasting of intuition opens the way to rethinking the relation between time and space, proposed by Kant under the rubric 'transcendental aesthetics'. Derrida reveals that the account of internal time consciousness presupposes the spacing of consciousness, as held in place between the parallel tracks of its empirical and its transcendental moments. He goes on to intimate that the differences between this spacing, or topology of consciousness, and the topographies of consciousness proposed by Freud are not of kind but of degree.

Différance is here introduced as interminable delay, *Verspätung*, but will subsequently be connected up to that other Freudian notion, *Nachträglichkeit*, the delayed impact of an affect, for which the mode of registration is invented after the fact. This provides *différance* with a structure of a doubled movement. In the essay '*Différance*' from 1968, of which there are conveniently two versions, both in English and in French, Derrida famously introduces the notion of the double meaning in the Latin of *differe*, of both a differing at a single moment, and a delay across a duration.[31]

> 'To differ' in this sense is to temporalize, to resort consciously or unconsciously to the temporal and temporalizing mediation of a detour that suspends the accomplishment or fulfilment of 'desire' or 'will' or carries desire or will out in a way that annuls or tempers their effect. We shall see later in what respect this temporalizing is also a temporalization and spacing, is space's becoming temporal and time's becoming-spatial, is 'primordial constitution' of space and time, as metaphysics or transcendental phenomenology would call it in the language that is here criticised and displaced.
>
> (D: 1968, SP p. 136)

There are then two further aspects of the thinking of *différance*, which get covered over in the emphasis usually put on its double structure as a structural figuration, in synchronous difference and diachronous deferral. This elides the further determinations of time, as historical condition, and of time, as duration of a genesis. Time as historical condition, and time as duration and genesis, are re-thought by Heidegger, Levinas and indeed by Derrida, to announce the arrival of the present, out of the future, with significant differences between them about how they conceive of that future. The figural distinction between synchrony and diachrony covers over differences between a thinking in which time is not made questionable, and a thinking in which it is made questionable: in a thinking for which genesis is the primary consideration.[32] However, the thinking which seeks to make time thematic is the one which in fact prevents a questioning of time, as stated in Augustine's paradox: 'When no-one asks, I know what it is, but when I am asked, I do not know' (*Confessions* XI).[33]

While Derrida's discussions of the notions of inscription and sedimentation from Husserl's 'The Origin of Geometry' have been taken to anticipate the Derridean notion of *écriture*, less attention has been given to an impact on Derrida of Husserl's questioning of Kant's distinction between a concern, in transcendental aesthetics, for an analysis of the forms of space and time, and, in a transcendental analytics, with a deployment of concepts. The challenge to Kant's rigorous distinction between the domains of intuition and of understanding is unpicked already by Husserl, who makes time into the dimension for the arrival of possible fulfilments of meaning intentions, and thinks space, both empirical and transcendental, as the medium for delimitating domains of regional ontologies, and for setting out the resulting horizons, within which meanings are deployed. The interdependence of these various constructs is perhaps best traced out in the relations between the two separately published parts of the 1907 lectures, *The Idea of Phenomenology* (HUA 2: 1950) and *Thing and Space: Lectures from 1907* (HUA 16: 1973), or again in separation of the first two parts of *Ideas: General Introduction to Phenomenology* (1913 and HUA 3, 4: 1976, 1952). Husserl opens out a possibility of thinking space, time and a deployment of meaning as inseparably bound up together, and of taking a further step, of supposing that, as configurations of space and time shift, so do understandings of how meaning works. This shakes the basis for Kant's distinction between a transcendental aesthetic, concerned with the pure forms of intuition, and a transcendental logic, articulating together the evidence of sensation, through the application of the categories in judgment.

The presumption that these two parts of the analysis can be brought back together to provide a single determinate set of concepts for analysing what there is in the world is subverted by Husserl's analyses, for which there is an inexplicable primordial insight, which constitutes, as a historically given *apriori*, the possibility of new conceptuality. The movement of ellipsis would then suggest that, when the new configuration of space, time and meaning arrives, the old one, with which it is not compatible, is not wholly erased, thus creating a disjunction, and a problem of thinking the transition between the two modes of configuring space, time and meaning. This might be thought to be the discovery of ethnography, as for example conducted by Lévi-Strauss among the Nambikwara.[34] I shall pursue the thought that Derrida's conceptions of *différance*, trace and iteration, and their successor terms, *restance*, *clandestination* and *destinerrance*, and the interruptive *achrony* of the *avenir* (future) as *a-venir*, to come, constitute a proposal to rewrite Kant's transcendental aesthetic, following up Husserl's proposed release of space and time from the delimitation imposed on it by Kant.

In Part II of these enquiries I pursue the conjunction of the names Husserl and Kant, and the less likely conjunction, Husserl and Freud, as they arrive in Derrida's readings of Husserl. Part III introduces the more obvious conjunction of the names Levinas and Heidegger. Through a series of engagements with the writings of each, in the course of which he develops the notion of a discontinuous series, the *sériature*, Derrida discards the hypothesis that the readings of Husserl, proposed by Emmanuel Levinas and by Martin Heidegger, might be construed as antinomies. In Part III Derrida's reasons for turning away from a Kantian diagnosis of antinomy, in favour of a Socratic diagnosis of an aporetics of reason are pivotal. This turn follows the move from constructing a framework of enquiry, as dating only from the beginning of the modern period, with, respectively, Descartes or Kant, to a framework dating back to a Greek origin of philosophy, with Aristotle and Socrates, Parmenides and Anaximander. Kant's antinomies, as introduced in the *Critique of Pure Reason*, concern, in terms of quantity, the possibility of determining a beginning and ending of time; in terms of quality, the finite or non-finite divisibility of a moment; in terms of relation, between an idea of freedom and an experience of determinism; and in the thinking of a possibility or necessity of necessary being.[35] Their role in the articulation of Kant's account of time, and the role of the antinomies in the *Second* and *Third Critiques*, modifying that account, is perhaps insufficiently noted.

The antinomies apparently rehearsed by Levinas and Heidegger are those of the priority of non-finite or finite time, and of the death of the other, or of a being-towards-death as a structure of Dasein. The discussion of the alternate trajectories of a possible impossibility, or impossible possibility of a conception of limit, introduces, as a third party to the discussion, the name Maurice Blanchot. I shall argue that these, while having the form of antinomies are not, strictly speaking, antinomies at all, since Levinas and Heidegger work with different conceptions of time, space, meaning, and indeed of the connections between them. Derrida's readings of Levinas and Heidegger, while starting in the early 1960s, continue in the texts from the early 1990s, *Gift of Death* (1992), *Aporias: Dying – Awaiting (One Another at) the 'Limits of Truth'* (1993), and *Specters of Marx: The State of the Debt, the Work of Mourning and the New International* (1993); and in the texts dedicated to Levinas and Heidegger separately, *Adieu to Emmanuel Levinas* (1996) and *Of Spirit: Heidegger and the Question* (1987). This makes any succinct assessment of his relation to each and to both together an impossible undertaking. Levinas, in a series of responses to Husserl, beginning with his involvement in the translation into French of *Cartesian Meditations* (HUA 1), in 1931, expands the domain of what must be thought to be given in advance of meaning intuitions and fulfilments. He expands on Husserl's notion of the pre-predicative, into an account of a priority of alterity, of infinity and of the arrival of the claim of the other. The horizontality of Husserl's delineation of the arrival of meaning is disrupted by an invocation of the height, from which a sense of the divine arrives, and the vector of intentionality is reversed, no longer moving from consciousness to its object, but from its divine origins to its mortal registration.

Husserl's enquiries enact an oscillation between a focus on an analysis of the results of the latter movement, in passive synthesis, and an analysis of the emergence, or genesis, of the former movement in active synthesis. Husserl himself introduces a doubled intentionality, in which an intending with respect to meaning contents is doubled with respect to an intending with respect to temporal ordering.[36] Heidegger, notoriously and as disputed by Husserl himself, seeks to ground the possibility of the data or evidence supporting their shared presumption of a priority of finite time, in a hypothesis and demonstration, of the meaning of being as given in and to Dasein, determinate being. This is the task undertaken but not completed in *Being and Time* (1927), the fragmentary nature of which must be a topic of discussion. Heidegger there proposes to demonstrate a derivation of intentionality and of the presentations of consciousness from the movements of time given in the ecstatic temporality of Dasein. This, however, was to be done in a section of *Being and Time* which was never published, and perhaps never written. Thus neither innovation, that of Heidegger, and that of Levinas, may be thought to be entirely successful, nor indeed complete. The impossibility of a givenness of meaning and evidence to support these hypotheses, except as pre-predicative, or older than time itself, or proleptic, never to arrive in the forms made available and constrained by an arrival of time determinations in languages, already marked by time determinations, should perhaps be unsurprising.

Derrida's readings of Levinas' texts explore the manner in which Levinas is compelled to re-enact the very violence of separation which Levinas diagnoses as characteristic of the Western tradition of philosophy. In 'Violence and Metaphysics: An Essay on the Thought of Emmanuel Levinas', Derrida traces how the use of the various tenses of the verb 'to be' carries with it commitments to temporal and ontological orders, which are strictly not available according to Levinas' own theorising.[37] At this point, another Kantian concept, that of transcendental illusion, becomes relevant, with an affirmation of Kant's hypothesis that it is structurally ineliminable from philosophical analysis, playing a role constitutive of the results of philosophy. This thought comes to expression already in *Of Grammatology*, in the tracing out of the effacement of the differentiations and temporal delays of signification, in the presumption of a self-presence of meaning in the voice:

> This experience of the effacement of the signifier in the voice is not merely one illusion among many – since it is the condition for the very idea of truth – but I shall elsewhere show in what it does delude itself. This illusion is the history of truth and it cannot be dissipated so quickly.
>
> (D: 1967, OG p. 20)

Derrida connects this back to the analysis of meaning in Heidegger's *Being and Time*, but it makes more sense once an account of Husserl's responses to Kant's diagnoses of transcendental illusion is in place. Derrida in effect adds a fourth source for transcendental illusion to the three diagnosed by Kant: the self, the world and the concept of God, which are all subject to the further transcendental illusion, inseparable from the concept of meaning: just because meaning is intended does not entail that there is determinate meaning. Responding to Kant on transcendental illusion brings Derrida's enquiries into close proximity with those of both Lyotard and of Foucault, on the history of truth.[38]

The text, *Speech and Phenomenon*, while overtly a study of Husserl, is perhaps also a response to the Kantian critical system, and to the absence in both of a theory of language, as a theory of meaning. For the emphasis on pointing up problems with Husserl's theory of signs in effect focuses on an aspect of Husserl's phenomenology of lesser importance to Husserl, rather more important for the neo-Kantian version of Husserl, provided by Merleau-Ponty, which separates out questions of meaning, from an analysis of signification and from analysis of ontological determination. This introduces a curious feature of Derrida's writings, whereby the overt address to one theoretical construct often masks an address to another, or to a conjunction of the first with the second. The readings of both Levinas and Heidegger are interrupted by invocations of Kant, especially in the address from 1980, 'Of an Apocalyptic Tone Recently Raised in Philosophy', in which Derrida mobilises Kantian critique, as a critical moment, with respect to both Levinas and Heidegger, in a rethinking of time and tone.[39] There is also detectable here a dispute between Christian and Jewish understandings of apocalypse, and a question to the effect on the Hebrew Bible of its transposition into the Christian *Old Testament*. There is also here a questioning of the impact on both of the invention of literature, at this point primarily in the terms of those forms of literature practised by Blanchot and James Joyce. This questioning of a relation between religion and philosophy, and between theology and literature is one of the concerns of the third and fourth parts of this book, where Derrida's encounters with the thinking of Blanchot and of Benjamin move to the centre of attention.

The puzzle of Heidegger's analyses of time, taking the place of those of Husserl, in the account of time problematised by Derrida in the essay '*Ousia* and *Gramme*: Note on a Note in *Being and Time*' has already been remarked. A reading of Heidegger also plays an important role in the 1967 text on Husserl, *Speech and Phenomenon*, in the diagnosis of a metaphysics of presence. In the essay, Derrida gives a reading of Heidegger's distinction between fallen, derived, or inauthentic time, and the authentic time of Dasein, also described as the distinction between 'improper' and 'proper' time, and he diagnoses as inevitable a re-installation of a metaphysical distinction between the two, re-enacting an erasure of a primordial time, as beyond such distinctions. This re-installation and re-enactment is diagnosed by Heidegger as characteristic of the Western tradition of philosophy. In subsequent writings, Heidegger seeks to escape this re-inscription by departing from the language of metaphysics in the direction of a more primordial thinking.[40] Derrida doubles Heidegger's conception of a metaphysics of presence, which he deploys against Husserl, with the notion of a metaphysics of the proper, which he deploys against Heidegger, not only in the writings of the 1960s, but still in the fourth *Geschlecht* paper, 'Heidegger's Ear: Of *Philopolemology*' (1989).[41] This metaphysics of the proper connects into a notion of an ex-appropriation, in which stable identities turn out to be dependent on the meanings and functions of the unstable identities, to which they appear to be opposed.

The temporal movements distinctive of the readings, in parallel, of Levinas and of Heidegger, and the series of doublings they generate, can be shown to be borrowed from Heidegger, rather than from Levinas. However, by the time of the work, central to this reading, *Aporias: Dying – Awaiting (One Another at) the 'Limits of Truth'* (1993), Derrida has arrived at a remarkable equipoise between the two responses to the phenomenology of Husserl, despite an initial prejudice in favour of Levinas and against Heidegger. Where Levinas can be understood to think in terms of a dissymmetry, generating a curvature of space and time, with the distance between humanity and the height of divinity not the same as the distance from divinity to humanity, the movement, characteristic of this reading, and constitutive, for Heidegger, of the relation between authentic and everyday time, is one of oscillation. Derrida's address to Levinas and to Heidegger moves back and forth across the disjunction of disagreements between Heidegger and Levinas, the better to engage with their thought. He does not follow Levinas into an enquiry concerning dissymmetry, as a determination of time. Where Levinas rejects ontology in favour of a metaphysics of the infinite, Heidegger delimits metaphysics as a worn-out phrase and phase of philosophy, seeking, instead, in *Being and Time*, to inaugurate a new departure in a fundamental ontology and, in *Of the Event: Contribution to Philosophy* (1936–38), developing an account of another beginning. This 'other beginning' is simultaneous with the foundation of philosophy among the Greeks, but repressed within it, permitting an alternate account of space, time, and meaning to emerge at the end of the evolution of the separation between *technics* and *poiesis*, which Heidegger supposes characterises the twentieth century. The event, the *Ereignis*, event as appropriation, is also for Heidegger the event of ex-appropriation, the *Ent-eignis*, in which what is given conceals its own origins. This *Ent-eignis* then lends itself by transposition into the notion of ex-appropriation, as deployed by Derrida, whereby that which is distinguished from any such other set of phenomena all the same is interdependent with those phenomena, while concealing their salience.

The movement from ellipsis to oscillation is a movement from the interruption, but interdependence of one line of enquiry by another, in favour of a deeper conception of interruption as *caesura*, a more emphatic break point, from which it is possible not to re-emerge. This presents a more complex relation between texts which pursue more strictly incompatible lines of enquiry, or indeed articulate incompatible lines of enquiry within what appears to be a single text. The juxtaposition of the incompatible accounts of time proffered by Levinas and Heidegger poses a question back to the coherence and sustainability of the account of time initiated by Husserl, within which Derrida inscribes his own account of time, by displacing the framing in terms of horizons and a temporalisation of consciousness, in favour of the thinking of *différance*, *restance* and the *a-venir*, the latter introduced in the essay 'Of an Apocalyptic Tone'. These determinations of time reveal themselves in the mode of registration and in the domain opened up as *écriture*, in the necessary movements of reading, imposed by texts themselves, as systems of meaning, and in the movements which texts impose on each other. Thus writing as textuality takes the place of Husserl's notions of pure grammar and of transcendental logic, as the framework stabilising meaning and making possible a limited deployment of concepts. Thus the thought of *différance* as a temporalisation of time emerges out of the readings of Husserl and Heidegger; while that of the *a-venir* emerges out of the readings of Levinas and Heidegger.[42]

The notion of *restance* focuses on that which cannot be recuperated within a theory of presence, or within a theory of identity as self-sameness, or within an account of time, which prioritises determinations of permanence, succession and co-existence. The latter are those put forward in Kant's analyses in the Analogies of Experience, of the *Critique of Pure Reason* (1781, 1787), and in their place, it is possible to advance these notions of interruption, of the *a-venir*, of *restance* and of *différance*, as introduced by Derrida, opening the way for a configuration of a series of alternate analogies of experience. Derrida re-deploys the concept of experience, beyond the limits imposed by a reading of history in terms of a metaphysics of presence. Experience is the double movement of a passage and an immobilisation, in a paradoxical arrival at a limit of meaning and sense. The notion of *restance* is explicitly introduced in *Glas*, or *The Death Knell: What Remains of Absolute Knowledge* (1974), the text in which Derrida releases Freud's conception of analysis, terminable and interminable, into Hegel's supposedly closed system of philosophy. Hegel's account of poetry and religion as partially achieved sketches of a philosophical completion is undone by mobilising the poetics of Jean Genet and the analyses by Karl Marx of the ideological functions of religion, into a reading of Hegel's *Philosophy of Right* (1821). The remains of the claims to absolute knowledge are thus one set of obscured sediments within the philosophical tradition to be explored; the remains of their Christian origins and commitments are to be traced out in the encounters between Derrida and Benjamin, in Part IV and between Derrida and de Man, in Part V.

The connections between a certain Eurocentrism and a certain privileging of Christianity within philosophy, as a consequence of the receptions of Greek philosophy in the name of Christianity, by Augustine, Thomas and Hegel, come to the fore through a reading of *The Other Heading: Reflections on Europe Today* (1991) and of *The Gift of Death* (1992, 1999), in which Derrida reads Levinas against Kierkegaard, thus complementing his own earlier mobilisation of Kierkegaard against Levinas in 'Violence and Metaphysics: An Essay on the Thought of Emmanuel Levinas' (1964, 1967).[43] Derrida's text 'Faith and Knowledge: Two Sources of "Religion" at the Limits of Reason Alone' (1994) provides a focus for attention, for the fourth part of these enquiries. It also provides a third occasion for considering Derrida's reading of Kant, alongside the discussion of the Idea in the Kantian sense, in Part II, and the challenge to the notion of antinomy, in Part III. In the discussion of religion and of Christianity, none of the parties, Kant, Levinas, Marion, and Nancy, presume the priority of an onto-theological delimitation of the scope and concern of religion, but are rather attuned to releasing religion from these constraints. Thus what is at issue here is not a re-determination, with Aristotle and Hegel, of the divine as life, continuity and eternal duration, nor any re-specification of the scholastic transcendentals of unity, uniqueness, completeness and truth. Rather, it becomes a question of the supposed separation between religion on one side and of theological concepts on the other. Levinas and Marion propose analyses of religion which abandon the categories of traditional onto-theology, while Marx and Benjamin re-affirm aspects of the classical theological conceptualisations, apparently independent of any religious commitment.

This re-opens a question of how to respond to the 'principle of all principles' of Husserl's phenomenology, with Marion disputing it, in a fashion markedly different from that of Derrida.[44] The 'principle of all principles' as stated in section 24 of *Ideas One* reads:

> But enough of such misdirected theorising. As principle of all principles, no such suspect theory can make us go wrong: that every originally given intuition (*Anschauung*) forms a justified source for knowledge, all of that which offers itself in 'intuition' (*Intuition*), that is in a living reality, is simply to be accepted as what it give itself out as, however, only within the limits within which it gives itself.
>
> (HUA 3, pp. 43–44)

Derrida takes this to prioritise an immediacy of givenness to sensory intuition, with which conceptualisation can never catch up. Marion seeks to persuade his readers that this givenness is indeed secured, but in the figure of a divine donation. The status, and interpretation to be given to this principle is thus keenly contested, re-opening a question to the connection from phenomenology to religious commitment, which is resolved differently by Husserl, Levinas, Derrida and Marion. The transition from the naturally given notion of *Anschauung*, to the more formally determined, but empirically less secure notion of *Intuition*, from the vernacular German to a Latin term also requires attention. The importance for Derrida of Nancy's conception of touch is that it provides a transcendental synthesis of the empirical faculties, in which such a givenness might be postulated, with neither the metaphysical commitments of the philosophical tradition nor the commitments to papal infallibility, which seem to follow along with Marion's philosophical undertakings. While Derrida's reservations with respect to Nancy on touch are rehearsed at length in *On Touching: Jean-Luc Nancy* (2000), there is some commonality on how to think syntheses of empirical evidences and meaning.

Derrida radicalises Nancy's, Heidegger's and Husserl's arguments for a priority of finite time, and for a thinking of the temporalisation of time as given within the movements of textual articulations. His hostility to the theology of the missed opportunity, articulated by Walter Benjamin, arises from this commitment to a radical finitude. The encounter with Benjamin, however, itself constitutes a missed opportunity to deepen the critique of religions, as the forces which bind communities, in favour of radical theology, divorced from its Christian partisanship. Derrida's readings of Augustine, of Nietzsche and of de Man show how strands of autobiography and of posthumous settlings of accounts leave open a problem for the securing of the posthumous life of the text, and for securing an anticipated memory of transience. This memory of inscription and inscription as memory are to be thought through an interweaving of Freud and Heidegger, on memory and forgetting, and of Husserl and Hegel, on retention, and on memory (*Erinnerung*) as interiorisation, and memory (*Gedächtnis*) as inscription. This generates a conception of inscription and of writing as always in some sense an exercise in self-constitution, retrieving the previously unregistered impact of previous inscription in the formation of the self, as the self who writes itself into determinate form. The contrast is with Benjamin who proposes a thinking of time, in which it is released from a conception of human history, as naturalised in the form of an eternal return, or as theologised in Hegel's conception of spirit. Benjamin's writings at all times emphasise what comes, briefly, and transiently to determinacy on the cusp of a movement from kernel to ruin, tracing out again the Aristotelian movement of generation and decay. Derrida in his return to reading de Man, in 1998, in 'Typewriter Ribbon: Limited Ink (2)' detects in de Man's writings a contrary movement, of releasing a notion of history as text from the movements of time. These contra-posed responses to a problem of thinking a relation between time and history, released from any naïve notion of progress, or any sectarian acceptance of a figure of divine providence, set out the context in which Derrida's account of bearing witness can be located, as independent of any doctrinal commitments or determinations of relations between religious and secular life, or between theological and philosophical concepts.

As remarked, in the course of these readings, Kant's determinations of time in the Analogies of Experience, as permanence, succession, and co-existence are disrupted and displaced through a rethinking of experience, which secures the internal coherence and cogency of the Derridean determinations of time, of temporalisation and of temporality as *différance*, *a-venir* and *restance*, each indicating a non-simultaneity of conceptual meaning, as transcendental, and any empirically given grasp of conceptual meaning, conditioned in the series of natural and historical time. A conception of experience is released from the stabilisations of metaphysical definitions, or as effected in Kant's critique, and in Husserl's phenomenology. Derrida shows how experience is better, more adequately understood as the repeated passage to various different kinds of limit, conceptual, kinetic, and kinaesthetic, and as the stalling of such passage, in a turning back from the limit. This provides a framing for thinking differences between the conceptions of experience, respectively endorsed by Husserl and by Kant, and the critiques of concepts of experience, offered by Levinas and Heidegger. In place of the transcendental unity of apperception which grounds the work of synthesis in Kant's critical system, and in place of the conceptions of an absolute, a transcendental and an empirical consciousness, in the relays of which Husserlian active and passive synthesis are held in place, Derridean concepts are held in place by a strong notion of textuality as autobiography, written into the margins of all those texts, held together by the marking of a single signature, as the writing into existence of a coherent trajectory of meaning.

The mobilisation of the workings of a Freudian unconscious, marked up in the terms *clandestination*, the obscuring of the ends of thought, and *destinerrance*, the vicissitudes of destiny, responds to and interrupts the supposed inexorable movements of Heidegger's historial framings and stabilisations of meaning, as well as disrupting Husserl's attempts to stabilise meaning in an absolute stream of consciousness, which Husserl supposes to have the form of time itself. The latter conception presumes that there must be a single outcome for all rigorously conducted conceptual enquiry; the former imposes a single framing on historical epochs, which suppresses whole swathes of human endeavour, in the focus on specific forms of passage to the limits of conceptualisation. Derrida's readings of Husserl through Freud, and of Heidegger, through Levinas, reveal a multiplication of temporal streamings in the supposed singleness of Husserl's absolute consciousness, and they reveal Heidegger's understanding of a closure of metaphysics to be just one of a number of possible framings of a specifically Greek inheritance, which might not have been subjected to a Christian domestication, nor yet to the Cartesian revision, and certainly need not be taken to be the only possible source for an originary impression for the movements of infinitisation and formalisation.

Interruption is the determination of time and the textual strategy by which Derrida disrupts an illicit importation into Husserl's thought of Kant's category of permanence. Interruption, as both textual strategy and as feature of temporalisation, permits a rethinking of an asymmetry between past and future which allows the determinations of time in the tense structure of the language of analysis to be used at the same time as putting the underlying conception of time into question. *Restance* is a determination of time allowing for a residue of preceding thinking, which is not yet capable of mounting a challenge to current over-simplifications, to be retained alongside those current simplifications, with which it is incompatible, as a potential source of disruption to them. It also permits a non-teleological account of the development of thinking to be put in place of the simplifications of supposing that there is one internally coherent theoretical account holding in place the trials and errors of actual enquiry. An elliptical movement is set up, through an oscillation between the temporalities of interruption and the temporalities of *restance*, and it is this movement which characterises both Derrida's mode of reading texts and the temporalising of time, emerging out of, and underpinning his readings. An account of Derrida on time then necessarily goes by way of a response to his readings of the texts of the tradition.

1.2 'Aphorism countertime'

> 13: Conversely, no contretemps, no aphorism without the promise of a now in common, without the pledge, the vow of synchrony, the desired sharing of a living present. In order that the sharing may be desired, must it not first be given, glimpsed, apprehended? But this sharing is just another name for aphorism.
>
> 14: This aphoristic series crosses over another one. Because it traces, aphorism *lives on*, it lives much longer than its present and it lives longer than life. Death sentence. It gives and carries death, but in order to make a decision thus on a sentence of death, it suspends death, it stops it once more.
>
> 15: There would not be any contretemps, nor any anachrony, if the separation between monads only disjoined interiorities. Contretemps is produced at the intersection between interior experience ('the phenomenology of internal time consciousness' or space consciousness) and its chronological or topographical marks, those which are said to be 'objective', 'in the world'. There would not be any series otherwise, without the possibility of its marked spacing, with its social conventions and the history of its codes, with its fictions and its simulacra, with its dates. With so-called proper names.
>
> (D: 1986, AC p. 421)[1]

This sequence of aphorisms stages the surmise that, if temporality is thematised as a feature of internal time consciousness, then it must also be inscribed in a topography, charting its worldly instantiation. The relation implied in this topography, presuming an inseparability of place, *topos*, from its graphic inscription, commits Derrida to seek instruction on place, and its staging of a relation between time and space, in the writings of the traditions available to him.[2] Thus Derrida juxtaposes Husserl's notions of internal time consciousness and of a living present, with a practice of aphoristic self-containment and its correlate, the series, in which aphorisms are usually organised, and, intriguingly, numerically ordered. For Husserl, there would be a relation of subordination between worldly topography and the ideality of transcendental topology, thus permitting him to engage in theorising irrespective of the specificities of the linguistic medium in which he works. For Derrida, the worldly instance of topography has precedence. Time constituted in its inner unity is for Derrida always a worldly time, whereas for Husserl the worldly is pre-inscribed in its ideal possibility, which nevertheless is given nowhere other than in worldly experience. This pre-inscription is then disputed by Derrida, who thinks the two as parallel, with the transcendental inscribed only within its worldly emanations, and inconceivable without them. In the course of Derrida's readings of Husserl, a gap opens up between Kant's notion of transcendental philosophy and the rather different notion of the transcendental, to be found in Husserl's phenomenology. A further gap opens up between the Kantian notion of a world as split between secular, or causal, and sacred, or revelatory instances, in terms of which the scope of human duty and freedom can be articulated, and the Husserlian notion of world which is not thus split between a determination of causal series, and a sphere of human freedom, duty and responsibility.

The citation at the head of this section, aphorisms thirteen to fifteen, is taken from an essay on *Romeo and Juliet*, published by Derrida in 1986. The invocation of a 'death sentence' marks up a connection to the discussion, in 'Living On – Border Lines' (1978), of Blanchot's narrative, '*L'arrêt de mort*', and the fleeting reference to Husserl's phenomenology, to *The Phenomenology of Internal Time Consciousness* (1928), reveals both how the thematics of time and a questioning of phenomenology remain twisted together for Derrida, and how the possibility of meaning for Derrida never stops raising the problems, first analysed by him in relation to Husserl's texts.[3] The connections between time, the possible communicability of meanings, and an interiority of subjectivity, here bounded as a monad, indicate a further suppressed reference to Husserl's *Cartesian Meditations* (1931) in which Husserl redeploys Leibniz's notion of the monad, to describe the structures of inter-subjectivity.[4] The *contretemps* of the title is the accident, or mishap, of chance, and, as Derek Attridge points out in his brief introduction, it is both the counter-time of musical composition, and the looser, common sense notion of the inopportune. However, marking a difference between an adventitious mishap, the inopportune, and a significant misadventure reintroduces a kind of hierarchy of significance, which it would appear Derrida is precisely seeking to subvert. Thus, analyses of chance can replicate the very difficulty which Derrida identifies in Heidegger's and Husserl's writings on time, of surreptitiously reinstalling a privilege between two versions of the one structure, chance or time, when the enquiry requires leaving the relation between the two undecided, even to the extent of refusing the thought that there are two contrastable instances at all. The chance of chance turning out to be significant or insignificant is for Derrida no more amenable to a rule of hierarchical organisation than are the various notions of time, called authentic, inauthentic, primordial and derived.

The deferral of death in the death sentence, the accident and necessity of chance, and the *contretemps* of time become closely bound up together in the conjunction of the names Husserl and Freud. The emphasis on the determination of time as *contretemps* reveals a notion of bad timing which is resistant to Husserl's attempts to thematise time, by reducing it to the series of is appearances in an absolute flow of consciousness. Thus Derrida substitutes the notion of the absolute as passage for Husserl's notion of time as absolute flow, and this passage is one from one state of incomplete conceptualisation to another. Derrida supposes that a bracketing of the natural attitude is needed, since it reveals the erroneous and unreliable nature of common sense notions of time, but he is not convinced that time as continuity and flow can be reinstalled at the transcendental level. He arrives at the view, through his readings of Husserl, that time is irreducibly aporetic, and can be studied only by attending to the paradoxes of its specific articulations in language and text, rather than constructed, or, as Husserl would say, constituted, in the interiority of consciousness. This constitutes Derrida's paradoxical affirmation and departure from Husserl's phenomenology. The relation is not unlike Husserl's own appropriation of themes from Leibniz, and in particular of the concept or construct of a monad, averted to here.

Husserl's appropriation of Leibniz's concept of the monad in *Cartesian Meditations* (1931) puts considerable pressure on that notion. For it denotes for Leibniz entities which are wholly independent of one another, without parts, extension or figure, but with varying degrees of perceptual power. Husserl transforms this into a designation for the constitution of delimited spheres of meaning, subtended by the activities of a single set of intentionalities, but the apparent separation between subjectivities is complicated by Husserl's discovery of inter-subjective and inter-generational strands in any instance of subjectivity, thus setting up a contrast between Leibniz's and Husserl's deployment of the term 'monad'. This is further complicated by Husserl's unsuccessful attempts to write a Sixth Meditation, with which to conclude the homage to and revision of Cartesianism, only partly presented in the text *Cartesian Meditations*, which first appeared in the French translation of Levinas and Pfeifer, in 1931. This Sixth Meditation was subsequently readied for publication by Eugen Fink, in the 1930s.[5] This reintroduces an emphasis on an absolute layer of consciousness, as mediating between layers of empirical and transcendental consciousness. This supposedly had Husserl's approval, thus generating differences between Husserlians: those who emphasise notions disruptive of the circumscription accomplished in terms of absolute consciousness: pre-predicative experience, passive synthesis, and inter-subjectivity, for example Landgrebe, Zahavi and Steinbock, and those who seek to make connections between Husserl, Hegel, and an absolute idealism, for example Hyppolite, Fink and, in a different way, Sartre. The completions of Husserl's thinking proposed respectively by Stein, in her edition of the time consciousness lectures, by Landgrebe, in relation to the question of pre-predicative experience, and by Fink, with respect to an absolute layer of consciousness and a transcendental methodology, are not entirely compatible with one another, and generate alternate possible lines of enquiry, all attributable to Husserl. Thus the name 'Husserl' ceases to pick out a single internally consistent series of themes and arguments. It is in this context that I propose to read Derrida's texts on Husserl.

In 1953, Derrida began drafting what became the text, first published in 1990, *The Problem of Genesis in Husserl's Philosophy*.[6] At that time, the principal available published text by Husserl on time was *The Phenomenology of Internal Time*

Consciousness, published in 1928, as though edited by Martin Heidegger. However, this text was in fact put together in 1917–18, by Edith Stein, Husserl's assistant, from Husserl's lecture notes dating back to 1904/05. Since Stein was inclined to a realist reading of Husserl's phenomenology, she was therefore, perhaps, inclined to underplay the deeply paradoxical nature of Husserl's claims with respect to inner time consciousness, as the primordial form of time. This text has since been supplanted, in 1966, by a volume of the Husserl edition (HUA 10), *On the Phenomenology of the Consciousness of Internal Time* (1966), which gives the same material, supplemented by an attempt to attend to temporal layerings in the development of Husserl's thinking of temporality.[7] This has advantages, for the writings brought together in the 1928 text span the inauguration of the transcendental turn in Husserl, in 1907, and the introduction of the procedure of bracketing out the common sense ontological commitments of the natural attitude, which has especially significant effects on the thinking of time. It also points up a *contretemps* of dating the enquiries as either 1904/05, or 1928, or 1966. In this volume, there emerges the notion of an absolute time of standing streaming flows, for which, as Husserl says under the titles 'The temporally constitutive flux as absolute subjectivity', and 'The time constituting flow as absolute subjectivity', from the two available translations of HUA 10, section 36, 'we have no names', or 'we lack names'. (HUA 10, p. 75, Churchill p. 100, Brough p. 79). (Quoted D: 1962, HOG p. 82). This Husserliana text is in turn supplemented in 2001, by the publication of HUA 33, *The Bernauer Manuscripts,* from 1917/18, in which the emphasis in the first two texts on retentions into the past of present awareness, is complemented by equal attention to the work of protention, projecting expectations into the future.[8]

This questioning of how to think about the future is taken up emphatically by Heidegger, and indeed by Levinas and Derrida. According to Heidegger, Husserl has missed an important clue with respect to time, derived from attending to the future: the finitude given in the being-towards-death of Dasein. Levinas by contrast insists on the irrecoverable nature of the past, out of which present understandings of futurity arrive. For Husserl, understanding depends on preserving the immediately preceding terms, or sounds, in a string of terms or sounds, for a complete string or sound structure to be perceived. This pattern is then projected into the future, for confirmation or disconfirmation, that the string is, for example, an English sentence, or that the sound series is, for example, the opening of the *Marseillaise.* The difference between thinking of meaning in terms of sentences and thinking of meaning as following the structure of musical themes turns out to be helpful, since the latter, with the relation between melody and harmonics, permits a thinking of a relation between horizontal and vertical structure, and between movements taking place within these distinctions and those taking place across these distinctions. Levinas emphasises the differences between meanings which arrive from on high and those which arrive horizontally; Derrida emphasises differences of pace and rhythm, to which he draws attention by the pacing of his own writings. This opens up differences concerning the tempo of time, as steady and constant, or as open to acceleration and deceleration, as well as questions about the direction of time, as arriving out of a future into the present, or out of the past in the present, or out of both simultaneously, evenly or unevenly. Readings which suppose that temporality is given as absolute consciousness tend to suppress the paradoxes and difficulties of the analyses of time, in favour of addressing other more tractable questions.

The text of Husserl's *Ideas Two*, with its emphasis on distinct tasks of constitution, with respect to a world of the inorganic, a world of the organic, and to a world embued with thought, was published in 1952, in the newly founded Husserliana edition.[9] It had, however, already had an impact on Merleau-Ponty's discussion in *The Phenomenology of Perception* (1945), since Merleau-Ponty had access to a pre-publication manuscript, and is similarly mentioned as an unpublished manuscript in section 10 of Heidegger's 1925 lectures, *History of the Concept of Time* (1985).[10] Merleau-Ponty privileges analyses, which remain committed to an 'in the world' context for analysis, with the focus on the embodied status of cognitive processes, by contrast to Derrida, who seeks to place his engagement with phenomenology in the neutral in-between, subtended by the analysis of eidetic consciousness of *Ideas One* (1913), on one side, and the analyses of embodied consciousness of *Ideas Two*. The *contretemps* here then is the availability of pre-publication access to *Ideas Two*, and indeed of the other papers in the Husserl Archive. In a review of HUA 9, translated into English as *Phenomenological Psychology* (1962), Derrida emphasises the importance for Husserl, and for his own reading of Husserl, of the carefully maintained parallelism between the results of phenomenological psychology, and those of transcendental phenomenology, while insisting on the differences between their respective ontological commitments and their implicit temporalities.[11] This supports the line of Derrida's reading against those of both Fink and Merleau-Ponty, who emphasise one or the other strand.

While phenomenological psychology attributes empirically given life spans to its subjects, as bearers of thoughts and of cognitive processes, transcendental phenomenology introduces the notion of transcendental life, unconstrained by human finitude, as providing the horizon holding meaning intentions and intuitive fulfilments in relation to one another. As both empirical and transcendental instance, human beings are held in place by both of these quite distinct temporalities: that of empirical life spans and that of transcendental time, as inter-generational aspects of life. However, the manner in which access to both, and a mode of living between the two, may be conceived poses problems. Heidegger mistakenly presumed that his solution to this problem, in their formal unity as Dasein's being-towards-death, would be approved by Husserl. Derrida's contribution here is to identify Husserl's conception of transcendental life as indistinguishable from a conception of transcendental death, both to be read through the tendency, in the death drive of Freudian theory, to re-impose an inorganic stasis on living organisms. The introduction here of the name of Freud and of his analyses has a startling and interruptive effect on the flow of Husserl's analyses, and of Derrida's reading of them. It results in the postulation of a conception of death in life, and of life in death: *la vie la mort,* and of the notion of the *survivre,* the living on in which an intensification of life is figured as equivalent to death.[12]

In *Speech and Phenomenon,* the use of the name 'Freud' to mark up the arrival of a non-Husserlian account of time, and the use of the name 'Saussure', for a non-Husserlian analysis of the sign, mark the moment when the text definitively ceases to be a reading of Husserl and is instead a demarcation of a critique.[13] As Derrida there writes:

> Going *through* the First Investigation we must try to ascertain how far these concepts respect the relations between signs in general (indicative as well as expressive) and presence in general. When we say *through* Husserl's text, we mean a reading that can be neither simple commentary nor simple interpretation.
>
> (D: 1967, SP p. 88)

In that text, Derrida supposes that the later developments of Husserl's thinking in terms of the transcendental turn and the thinking of time are already implicitly present in the early text *Logical Investigations,* a claim disputed in various ways by various more convention-bound readers of Husserl. Derrida's insistence on reading across Husserl's texts, for a meaning held in place across Husserl's own discussions, thus deploys Husserl's own conception of an ideality of meaning, this one determining the meaning of Husserl's phenomenology, as a whole. This then conflicts with readings of Husserl which suppose that the turn to transcendental phenomenology and even more that which analyses historicity and inter-generational transmission make no sense, and must be resisted, in order to rescue what is valuable in Husserl's enquiries.

Against the notion of transcendental life, and its supposed equivalent, transcendental death, Derrida develops the volatile notion of the living on, the *survivre,* survival as an intensification of living, of that which overflows the limits of empirical life and death, but which is held in place by its connection to empirical life and death. The question to be asked here is whether the notion of *survivre* is more strictly Husserlian than the notion of transcendental life, which exceeds the evidences of primordial impressions, and indeed lends itself to an appropriation by Fink, in the direction of Hegel's concept of spirit. The notion of the *survivre,* intensified living, is developed by Derrida in his analysis, in *The Post Card: From Socrates to Freud and Beyond* (1980), of an interdependence, for Freud, of notions of life and death, in the reading of Freud's essay 'Beyond the Pleasure Principle' (1920). Thus, where Eugen Fink looks to Hegel's notion of spirit to gloss the notion of absolute consciousness, holding together the evidences of empirical and of transcendental consciousness, Derrida looks to Freud to elaborate a syncopation and lack of fit between the workings of the unconscious, the pre-conscious and consciousness, to articulate this same relation. Thus a Hegelian mediation of difference, by moving to a higher level of analysis, in the dialectical movement called sublation (*Aufhebung*) is contrasted by Derrida to the Freudian *Aufschub,* the deferral, of fulfilment of sexual drives in a sublimation.[14]

Freud remarks in his essay 'On Narcissism: An Introduction' (1914):

> the formation of an ego ideal and sublimation are quite differently related to the causation of neurosis. As we have learnt, the formation of an ideal heightens the demands of the ego, and is the most powerful factor favouring repression; sublimation is a way out, a way by which those demands can be met without involving repression.
>
> (PFL 11, p. 89)[15]

The essay concludes with the following observation concerning an instability in the distinction between ideal-formation and sublimation:

> The frequent causation of paranoia by an injury to the ego, by a frustration of satisfaction within the sphere of the ego ideal, is thus made more intelligible, as is the convergence of ideal-formation and sublimation of the ego ideal, as well as the involution of sublimations and the possible transformation of ideals in paraphrenic disorders.
>
> (PFL 11, p. 97)

Freud is acutely aware of the instability of his own conceptual distinctions and of the difficulty of describing the interaction between the two distinct series of connections: the thing connections of libidinal affect and the word connections of conscious articulation. Derrida's readings of Freud and of Husserl suggest a parallel between this problem of articulation and that of describing the process of assigning conceptual determinacy to pre-predicative originary impressions. There are problems in parallel between describing the relation between thing-connections and word-connections, and between the empirical and the transcendental processes in play in each of these: originary impression, conceptual determination and the process of transposing the one into the other.

The irreducible residues which according to Freud resist sublimation (*Sublimierung*) and articulation into language subvert the Hegelian movement of a sublation (*Aufhebung*) without remainder. This motivates the introduction of the term *restance*. Freud's notion of transference (*Übertragung*) is deployed by Derrida in the margin of the text 'Living On – Border Lines' (1978) to bring into question the possibility of any translation of meaning (*Übersetzung*) from one language to another. By contrast to the hypothesis that natural languages can be thought to approximate to one and the same set of ideal meanings, the movements of transference and its counter-movement, counter-transference, subvert in advance the thought that there is a determinate text given in advance. Instead, there are two separate processes of meaning intention, in the transference and in the counter-transference, which are in conflict with each other for hegemony. Analogously, there are two separate under-determined sequences of meaning twisted together in any attempt at translation: the current activity of producing meaning, and that of the partially pre-given text. There is then a process of sedimentation of meaning, with the first text, to be translated, or narrated in the analytical session, becoming inextricably intertwined with the meaning intentions at work in the current context, or process of attempted translation, into which they are brought. A parallel process might be thought to take place in relation to Husserl's analyses, which, in attempting to arrive at a level of lucidity through reduction to the level of transcendental constitution, nevertheless must always interrupt themselves and be re-inscribed in actually given empirical awareness. The point of Husserl's notion of the living present as a transcendental structure then is to reveal the immediacy of empirical awareness as embedded in transcendental and trans-temporal conditions, such that these transcendental conditions are given simultaneously with empirical givenness, but mediately and as adumbrations, not fully given as complete components of that awareness. The processes of phenomenological analysis then are, for Derrida, like the processes of a psychoanalytical analysis: always at risk from an interruption and prone to wander from one goal of analysis or description to another, as a result of the emergence of some other further preliminary consideration. Derrida introduces the notions of *destinerrance* to name a movement of an unavoidable wandering away from any supposed origin, in a series of deviations revealing an instability of the origin itself; and a *clandestination* which obscures any thought of a *terminus ad quem,* of a determinate end point of the process. There are here neither determinate points of beginning nor determinate end points.

This diagnosis of a problem for Husserlian analysis is brought into focus by a series of reflections Derrida makes on the notion of chance, and on marking the arrival of the unexpected, in his essay '*Mes Chances*: A Rendezvous with Some Epicurean Stereophonies' from 1982.[16] This essay poses questions most directly to a connection between Freud and Heidegger's analyses of time and movement, but can also be deployed to open out a question to contrasting directions of movement in Heidegger's analyses and those of Levinas. By extension this essay thus also touches on the thinking of time and movement in the writings of Husserl and indeed Aristotle. Derrida writes there:

> For the time being, let us be content to take note of this law or coincidence, which in an odd way associates chance and luck with a descending movement, a finite throw (which is supposed therefore to fall vertically again), the fall, the incident, the accident, and most certainly, the coincidence. The attempt to submit chance to thought implies in the first place an interest in the *experience* (I emphasize this word) of that which happens unexpectedly. Indeed there are those of us who are inclined to think that unexpectability conditions the very structure of an event. Would an event that can be anticipated and therefore apprehended or comprehended, or one without an element of absolute encounter, actually be an event in the full sense of the word?
>
> (D: 1982, MC p. 5)

To this line of questioning he responds:

> There are those who lean toward the assumption that an event worthy of the name cannot be foretold. We are not supposed to see it coming. If what comes and then stands out horizontally on a horizon can be anticipated then there is no pure event. No horizon, then, for the event or encounter, but only verticality and the unforeseeable. The alterity of the other – that which does not reduce itself to the economy of our horizon – always comes to us from above indeed, from the above.
>
> (D: 1982, MC p. 6)

The downward throw is twinned by the arrival from on high, of a Levinasian height, an 'above' from which the unexpected arrives, and puts a question to the capacity of the phenomenological notion of the horizon to make room for the moment of the arrival of the as yet unthought. This suggests that a horizon conceived in accordance with the topography of the spherical earth has to be denaturalised in order to permit the Husserlian thought of the horizon to arrive.

An empirical notion of the horizon precludes the arrival of the unknown, as Derrida supposes, but a transcendental conception of the horizon is not bound up to the metaphorics of visibility, invisibility and pre-visibility, and is not inscribed within a naturalised conception of time. The metaphorics of the horizon appear to reduce time to a spatialisation, but for a transcendental horizon, time and space are not mutually opposed notions but, precisely as envisioned with the term *différance*, provide a thinking of space-time coordinates through which an historical *apriori* becomes genuinely thinkable. For such an *apriori*, then, there is a time in history when that *apriori* becomes thinkable for the first time, which Husserl calls its *Erstmaligkeit*, but which then has the status of an omni-temporality, such that it pre-dates the time of its own first formulation. This in turn becomes thinkable and makes sense only once a distinction between empirical and transcendental, or between natural and an eidetic time has been worked out. Thus there is here a distinction between empirical and transcendental horizons, and the historical *apriori*, for which the empirical and the transcendental are inseparably intertwined such as to constitute this paradoxical temporality. This may prevent Derrida's objections to Heidegger, on distinguishing authentic and inauthentic time, from also posing problems for these more Husserlian distinctions. Pursuing this line of argument requires a transformation of Derrida's own notion of *différance*, from one apparently opposed to Husserl's horizonality, to one which, through the disruptions of the thinking of the future, as the *a-venir*, become a retrieval and deepening, indeed a rebirth of Husserl's thinking of the horizon and of the historical *apriori*.

Paul Ricoeur, in his introduction to his translation into French of *Ideas One* (1950), provides a concise observation concerning another of the central concepts of Husserl's phenomenology, intentionality, illuminating the relation of parallelism between the results of empirical and of transcendental phenomenological enquiry.

> This is why intentionality can be described both before and after the phenomenological reduction; before, it is an encounter; after, it is a constitution. And it continues to be the common theme of pre-phenomenological psychology and of transcendental phenomenology. Reduction is the first free act because it is the one that liberates me from mundane illusion. Through it I apparently lose the world that I truly gain.
>
> (R: HP, p. 21)[17]

In his notes to his translation, Ricoeur marks up various features of the reception of Husserl at this time, not least the one-sidedness of Merleau-Ponty's insistence on analysing consciousness, as it is given in the world, leaving out the analysis of reduced consciousness. This remains a reference point for Derrida's reading of Husserl from his *Introduction to Husserl's 'Origin of Geometry'* (1962) to *On Touching: Jean-Luc Nancy* (2000). Ricoeur also marks up a certain hesitation with respect to Eugen Fink's claims to present a version of phenomenology indistinguishable from Husserl's own, while praising Fink's attempts to distinguish Husserl's transcendental commitments, from those of Kant. This hesitation is perhaps insufficiently marked in Derrida's readings of Husserl. Ricoeur's translation of *Ideas One* took its place alongside the translation of *Cartesian Meditations* (1931), undertaken by Pfeifer and Levinas, also from a text not yet established by the scholars at the Husserl Archive, Leuven. There are thus differences between the Husserl texts used for these translations, and those subsequently established in Husserliana editions, and Derrida's readings are irretrievably marked by the effects of this gap and by the receptions given in France to *Ideas One*, in Ricoeur's translation, to *Ideas Two*, by Merleau-Ponty's use of it in *Phenomenology of Perception*, and by the responses to Husserl of Emmanuel Levinas, first through his part in the translation of *Cartesian Meditations* and then through its impact on his philosophical formation, especially the lectures *Time and the Other* (1948).

These lectures identify the interdependence of problems in Husserl's analyses of time, and of association and passive synthesis, and problems in his analyses of consciousness, inter-subjectivity, and of otherness.[18] The conjunction 'time and the other' makes its mark on Derrida's reading of Husserl in *Speech and Phenomenon* (1967). It is, however, Ricoeur, who in an essay from 1949, 'Husserl and the Sense of History', identifies in Husserl's *The Crisis of the European Sciences and Transcendental Philosophy* (1936) Husserl's reliance on a version of the 'Idea in the Kantian sense'. Ricoeur identifies in the deployment of this idea two conflicting movements to the limit:

> Philosophy is the 'innate entelechy' of Europe, the 'proto-phenomenon' of its culture. Indeed, to be European is less a glory which particularizes than a responsibility which relates to all. Again, it is necessary fully to understand this term: philosophy. Understood as the sense of European man, it is not a system, as school, or a work with a date, but an Idea in the Kantian sense of the term; it is a task. The Idea of philosophy, this is the teleology of history.

> This is why the philosophy of history, in the end is the history of philosophy, itself indistinguishable from philosophy's coming to self-awareness.
>
> (R: HP, pp. 152–53)[19]

Derrida retains this appreciation of the importance for Husserl of the Idea in the Kantian sense, but he breaks with this concept of a unitary history. I shall return to this in discussion of Derrida's *Introduction to Husserl's 'Origin of Geometry'*, in the next chapter.

Ricoeur indicates a tension between conflicting concepts of philosophy, as achieved totality and as infinite task:

> But, what is philosophy as an Idea, as a task? What is its relation to the whole of civilization? From the start, to designate philosophy as an Idea is to emphasize its two traits of totality and infinity. Husserl even calls it a telos, an end aimed at, for it is the telos of the science of the whole of being. Because it is directed toward the achievement of the science of all that is, the Idea of philosophy can be only a 'normative form situated at infinity', a pole at 'infinity'. Each historical realization of philosophy still has the inaccessible Idea for its horizon.
>
> (R: HP, p. 153)

Nevertheless, there is still an underlying unification here in the notions of horizon and of a 'telos of the science of the whole of being'. The recurrence of the two terms 'totality and infinity' as opposed, in the title of Levinas' text *Totality and Infinity: An Essay on Exteriority* (1961) is not fortuitous, and Levinas disputes the thought that they may be conjoined in an affirmation of a completable system. Levinas and Ricoeur both obliquely invoke the discussion of these terms by Kant, in the first two of the Antinomies of Reason, concerning quantity and quality, from the dialectic of the *Critique of Pure Reason*, which explores the internal instability of each.[20] Levinas' emphatic proposal to recast a thinking of infinity throws into question the unity of any philosophical trajectory, marking a serious departure both from Kant's critical system and from the unifications of Husserl's thinking. The result is a reversal of the direction of intentionality, which places priority on a divine time of beginning, in a past which was never present, and places the thinkability of the thought content in advance of it arriving to be thought by human beings. This reversal of the direction of intentionality is not precluded by Husserl, but there is for Husserl a possible retrieval of that thinkability, in a determinate thought content, as a result of performing the required reductions. This both underlines and puts in question the effects of Husserl's notions of genesis and of passive synthesis, in which transcendental subjectivity is constituted, as a consequence of and through the processes of attending to the arrival of possible intended meanings. For these two, for Levinas, open out points of origin for meaning, beyond any retrievable horizonality.

The problem of positing an infinity, which is irrecuperable within a totality of possible thought contents, is replicated at the level of time constitution. The time of those possible meanings is formed in advance of any time of registration, and this time, in advance of its registration, is Levinas' immemorial time, one which cannot be brought into the horizon of the lived presence. This time then vies for precedence with the time of originary evidence, which is given in originary, if pre-predicative impressions, and which for Husserl must take precedence, as absolute flow. The past which was never present, the absolute past, is not retrievable into the time horizon of what can be presented or re-presented in presentification (*Vergegenwärtigung*). Thus Levinas' absolute past competes with Husserl's notion of an absolute flow, a contestation which could be understood to be resolved by Derrida's notion of the absolute as a necessary and impossible passage between the two. In relation to this hypothesised immemorial time, Derrida develops a number of different responses, one of which is affirm the proposition '*Tout autre est tout autre*', 'every other is wholly other', in the last section of *The Gift of Death* (1992).[21] The affirmation of an irreducible otherness, as an alterity in advance of sameness, destabilises the unifiability of a thinking of time, but it is not the case that Derrida wholly accepts the full implications of Levinas' critique of Husserl. Levinas' disruption of Husserl's account of time plays an increasingly important role, in the move from Derrida's readings of Husserl, from 1953–54, into the 1960s, and even more so when the phrase 'every other is wholly other' emerges as a way of twisting together the problems of thinking otherness and thinking time, of thinking infinity as otherness, and totality as sameness, in a grammar which has not presumed the very continuities of time to be put in question. This then is an explanation of the point and function of Derrida's neologisms: to hold open the possibility of such a non aligned grammar.

In the course of Derrida's reading of Husserl, there are three key Husserlian distinctions to which Derrida draws attention, and the status of which he disputes. The first is the distinction between immediate presentation and mediated presentification. The second and third are the supposed distinctions between reduced time and the cosmic time of worldly occurrences; and between static and genetic phenomenology. The challenge to the soundness and stability of these connected distinctions is crucial for Derrida's third text on Husserl, *Speech and Phenomenon: Introduction to the Problem of Signs in Husserl's Phenomenology* (1967). In each case, Derrida supposes that the first preferred term cannot be kept separate from the second. The connections between these three distinctions are also significant for Derrida, and for his developing view of Husserl's phenomenology. The first distinction turns on a further distinction between perception, as responding to immediate intuitive givenness, with a continuous attentiveness of impression, protention and retention, by contrast to the overcoming of discontinuities in awareness, in presentification. The contrast is between a continuous temporality, connecting present impression, protention, and retention, which Husserl at first calls primary memory, and the discontinuity bridged in secondary memory. However, Derrida's critique identifies how for Husserl the difference between these two must present itself as a primordial impression, given in a continuous present of the living present, thus subordinating discontinuity to continuity. For Kant and for Husserl continuous time is the more basic; for Levinas and for Derrida discontinuous time, or interruption, is the more basic. If discontinuity takes precedence over continuity, then secondary memory would take precedence over retention, over primary memory, with a *husteron/proteron.* However, Heidegger's thinking of the event, as the disjoining of continuous time, reveals that there is already a thinking of such discontinuity in Husserl's thought of historical *aprioris*, as the arrival of *apriori* conceptual formations, in the course of history. Euclid's notion of geometry, Plato's conception of the idea, and the related notion of infinitisation, and Husserl's concept of formalisation, by contrast to that of generalisation, might all be thought to be instances. *Différance* may be another.

The second distinction, between cosmic, or worldly time, and an eidetic time, the result of bracketing natural prejudices, and producing reduced phenomenological time, opens out a distinction between time as immediately experienced and the time of the living present, resulting from repeated performings of a bracketing of uncritically given conceptions of time. This raises the problem of how an intuition concerning the essential structure of temporality might be given, as either categorial or eidetic intuition. Since this distinction forms the basis for attempting to distinguish between a natural, or vulgar, concept of time, and its origin in an essential temporality, as source or temporalisation of time, it links into Derrida's discussion in '*Ousia* and *Gramme*' of Heidegger's distinction between derivative, or inauthentic, and an authentic, or ecstatic temporality. In phenomenological psychology, essential temporalisation appears inseparable from natural time, and is distinguishable in a transcendental phenomenology only once the phenomenological procedure of bracketing and reduction has been undertaken. In the domain of worldly experience, essential temporality appears as, and is inseparable from natural time order, but the more it is analysed, the less natural 'natural' time appears. For Derrida, as indeed for Husserl, the characterisation of a 'natural' time of linearity and punctuality can emerge only on the basis of the workings of a temporality of delay, curvature, and an elliptical movement back and forth between intending act and intended content. Husserl's version of this delay and swerve is to be found in the textures of his manuscripts, as patiently transcribed in the Archive.

This second distinction is further complicated by the discussion of Husserl's claim, in Derrida's *Introduction to Husserl's 'Origin of Geometry'* (1962), that even the most formal thought structures, those of geometry, have historical conditions of possibility. This marks up the important point that for Husserl, as indeed for Heidegger, there is a third notion of temporality, that of historicality, alongside those of the natural, or derivative, and the phenomenological, or authentic. This disrupts the supposed polarity between the first two, and complicates the claim of the essay, '*Ousia* and *Gramme*', that there is only one concept of time, and that it is the vulgar concept. A splitting between time and temporality emerges for attention. There are important differences between Husserl and Heidegger on the nature of this historical time, with perhaps a significant shift for Heidegger between 1927 and 1938. Husserl insists on an historicality of the originary intuition of the meanings of ideal essences, and on the need to reactivate meanings, to disrupt the sedimentations of that meaning resulting from its transmission as signitive, rather than as intuitive meaning contents. Thus transmission of theoretical contents from one generation to another is deemed prone to such sedimentation, embedding intuitive meaning in potentially distorting signitive transmissions. Derrida questions whether there is ever an intuitive meaning fulfilment entirely free of signitive elements, that is free from the mediation of language use and signification. Thus the distinction between primordial intuition and secondary signification is put under pressure.

The questioning of these distinctions is further complicated by Derrida's exploration of two further problems. There is the problem of translating one natural language into another, marked up with especial force by the occurrence in the one, but not in the other of homonyms, such as the play on words in French, but not in English, of *temps*/tense/time and *lettre*/letter/sign. There is also the problem of transposing from the level of immediate consciousness to that of its transcendental conditions of possibility, of which, in immediate consciousness, it is not possible to be conscious. Increasingly, in the course of the development of his responses to Husserl, from 1953 to 1967, and beyond, to the writing of *The Post Card: From Socrates to Freud and Beyond* (1980), Derrida appeals to the structures of Freudian analysis to articulate an unpresentability of these conditions, that is an in principle impossibility of presentation in consciousness of consciousness, which he supposes is covered over by Husserl's phenomenological commitments. The third distinction put in question by Derrida is that between a static constitution of time, given as a living present, and a genetic constitution of time, as given in the interaction between retention and protention on the one side, and between the fallibilities of secondary memory and secondary expectation on the other. For Husserl, there is the natural time of uncritically considered experience; there are the time structures of the processes of presentation in perception and of presentification through secondary memory, in which the articulation of phenomenological data consists, and there are, third, the distinctive temporalities of the discrete domains of entities, as revealed by phenomenological analysis: for example those of acts (*noeses*) and of thought contents (*noemata*), of inorganic and organic contents, and of those embued with life, which form the basis of lived experience. For Husserl, there are difficulties in linking these various times and temporalities up to each other, without generating an infinite regress. For an articulation of the relations between the first and second of these, Derrida is inclined, in the first text on Husserl, *The Problem of Genesis*, to apply the word 'dialectics', although not in a way connected to the strong Hegelian notion of dialectics, but more as an equivalent to a notion of diachrony. This use, shared with Gaston Bachelard in his text *Dialectics of Duration* (1950), is held over into the writing of *Speech and Phenomenon*.[22]

The ingenuity of Derrida's text, *The Problem of Genesis*, is that it deploys a series of distinctions between static and genetic considerations within Husserl's thought to reveal a trajectory of enquiry and a structure of repeating problems delimiting a single structure, within which the stages of Husserl's enquiries can be positioned as integrated parts. This releases a narrative of Husserl's development of phenomenology from the accident of historical succession into a necessary iteration constituting a single theoretical structure. To achieve this, Derrida re-deploys Husserl's own invocation of the Idea in the Kantian sense, with the addition of a more rigorous invocation of some of Kant's distinctive theoretical commitments, through which Kant's concept of an Idea in the Kantian sense may be distinguished from Husserl's rather looser deployment of it. In the 1963 essay '"Genesis and Structure" and Phenomenology', Derrida sums this up in the following way:

> It is the infinite opening of what is experienced which is designated at several moments of Husserlian analysis by reference to an *Idea in the Kantian sense*, that is, the irruption of the infinite into consciousness, which permits the unification of the temporal flux of consciousness just as it unifies the object and the world by anticipation, and despite an irreducible incompleteness. It is the strange *presence* of this Idea which permits every transition to the limit and the production of all exactitude.
>
> (D: 1967, WD p. 162)[23]

In *The Problem of Genesis*, Derrida explores the links from Husserl's use of this term back to Kant's analysis of the working of antinomy with respect to notions of limit and the unlimited, in the Analytic of Kant's *Critique of Pure Reason*.

In subsequent writings, as will emerge, Derrida reads the Idea in the Kantian sense as linked up to some of the more systematic aspects of the Kantian critical system, concerning religion and revelation, and considerations of the human and the divine view of human destiny, with which Husserl is not necessarily in agreement, but with which the Husserlian programme can be usefully brought into contact. This takes place both in *The Other Heading: Reflections on Europe Today* (1991) and in the later essay 'The World of the Enlightened to Come: Exception, Calculation and Sovereignty', appended to *Rogue States* (2003). While the title on its own could bear serious scrutiny and commentary, I shall pick out here the notion of the 'world of enlightenment to come', and attend less well to the three abstractions, exception, calculation, sovereignty. The 'to come', the *a-venir*, is the distinctive mode of thinking futurity at which Derrida has arrived by way of a long reflection on the apocalyptic tone of philosophy, identified by Kant, and on his differences with Emmanuel Levinas, on how to think the disruption of philosophy by religiousness. This will be discussed in Parts III and IV of this study. In that essay Derrida orchestrates one more encounter in an extended series of such encounters in his writings, between the phenomenology of Edmund Husserl and the critical philosophy of Immanuel Kant, under the heading '*Sauver l'honneur de la raison*': to rescue the honour of reason, as a response to the thought of Jean-François Lyotard.

In that essay Derrida considers the logical illogic of the organism which self-destructs in order to avoid the consequences of an invasion by alterity. This figure of auto-immunisation serves as a model for thinking about how intellectual life in general, and philosophy perhaps in particular, are prone to adopt a mode of self-annihilation, in order to avoid the challenge of exteriority, the possibility that there might be sources of reason not yet taken up into the self-affirming circuits of a would-be philosophical reasoning. Those who would save the honour of reason may be found thinking it better not to do philosophy at all than to allow the entry of the unacceptable outsider. Saving the honour of reason imposes a unity of versions of reason in the name of an unexamined 'we'. The phrase indicates an illusion, to which Lyotard has drawn attention in *The Differend: Phrases in Dispute* (1983) under the rubric; the transcendental illusion of the '*nous*', the 'us'.[24] Of this Derrida has written in his paper, 'Lyotard *et nous*', delivered at the commemorative conference on Lyotard's work at the Collège Internationale de Philosophie in 1998.[25] 'But who, we?', as Derrida asked already in 1968, in the address to the New York colloquium of October 1968, now known as the essay 'The Ends of Man'.[26] Lyotard writes of this version of a transcendental illusion in section 155 of *The Differend*:

> A single proper name whether singular or collective, designates an entity astride two heterogeneous situations. It is the property of proper names to receive such heterogeneities (nos. 80, 81). But it is not legitimate, it is even illusory, in the Kantian sense of a transcendental illusion, to suppose a subject-substance that would be both a 'subject of the uttering' (even though it is not the addressor in the prescriptive) and the permanence of a self (even though from one phrase to the next it leaps from one instance situation to another). Its proper name allows it to be pinpointed within a world of names, but not within a linking together of phrases coming from heterogeneous regimens and whose universes and the tensions exerted upon them are incommensurable with each other. The 'we' would be the vehicle of this transcendental illusion, half way between the rigid (constant) designator that the name is and the 'current' designator that the singular pronoun is. It is not surprising that, in the 'currentness' or 'actuality' of obligation, the 'we' that reputedly unites the obliged and the legislator is under threat of splitting.
>
> (Ly: D, p. 99)

With Lyotard, then, at this later stage, Derrida poses a question to an assumed collectivity, above all in philosophical analysis, but not only there as instances of transcendental illusion.

The first question to be posed is whether Husserl's reliance on the Idea in the Kantian sense is compatible with the rigorous interpretation of the 'principle of all principles', concerning primordial data for meaning. The second question to be posed is whether the Husserlian separation of a transcendental time of fulfilments of meaning intentions can be separated out from a natural time of historical process, and a natural time of an ordinary conception of the future. Derrida subsequently adds another dimension to the questioning of the distinction between the immediacy of perceptual givenness and signitive delay, by intimating that the immediacy of perceptual givenness, in which meaning is fulfilled, the 'now' of fulfilment, or of self-presence, is an ana-chronic moment, which must fall out of time. This then links into the analysis of the *contretemps*, in *Romeo and Juliet*, as the postponement of the fulfilment of a promise of a shared simultaneity. The analysis of the distinctions between natural, phenomenological, and historical time puts pressure on the possibility of revealing a pure essence of time, undisrupted by empirical specificities of history, while the relation between static and genetic constitution displaces the origin of meaning back into a pre-predicative origin, which remains to be retrieved in a future intuitive fulfilment. In place of a threefold structure of time, as past, present and future, as given in natural time, and in the tense structures of modern European verb conjugations, an alternate account of time begins to arrive. This 'other time' is at variance with the tense structures of the languages in which Derrida and Husserl write; and this poses huge problems for its expression and articulation within the grammars of natural language.

The problematic status of a language adequate to the demands of phenomenology and for the exposition of eidetic structure perplexes Derrida, and indeed Eugen Fink before him. In the *Introduction to Husserl's 'Origin of Geometry'*, Derrida invokes Fink's preoccupation with a transcendental *logos*, the possibility of transcendental language, and the requirements of a transcendental discursivity, in which to articulate the results of transcendental enquiry, to which he returns in the opening section of *Speech and Phenomenon*: 'and as Fink has well shown Husserl never raised the question of the transcendental logos, the inherited language in which phenomenology produces and exhibits the results of its reductive operations' (D: 1967, SP pp. 7–8). An arch-writing, *écriture*, is not thus constrained by natural tense structure, and thus cannot be transposed into the linguistic forms governed by natural tense structures. It is thus the secret sharer, or contraband, of given natural languages, and might be thought to fulfil the requirements invoked in the notion of a transcendental *logos*. A fourth distinction, put in question by Derrida, alongside the three already indicated, is then that between an order of transcendental sense, and its approximate expression in natural language. Derrida repeatedly questions the distinction drawn by Husserl between an order of transcendental meaning and the givenness of actual languages. He questions whether Husserl can be granted the presumption that natural languages tend at the limit to realise or accommodate themselves to the form of pure grammar, or the form of a transcendental logic, and whether the pure form of transcendental meaning or pure grammar can be kept secure from the naturalisations and localisations imposed by particular natural languages. There are similar questions to be posed about the relation between phenomenological time and natural time.

Phenomenological time is given along with natural time, but is completely distinct from it, above all with respect to naturalising, but inconsistent, assumptions concerning time resulting in determinations of time as the punctual 'now', the time series, and the cyclicity apparent in the various rotations of seasons and planets. For Husserl, any suggestion that phenomenological time retains the same structure as so-called natural time, of a present time, preceded by a past time, and succeeded by a future time, is to be resisted, and this is made a focus for analysis from 1904 onwards. For Derrida, the separability of these two is precisely in question: can there be a pure thinking of a structure of time, unconstrained and uninflected by local, naturalising tendencies at work in worldly consciousness and restricted linguistic competences? There are then three parallel structures to be interrogated: there is the parallelism between empirical psychology and transcendental phenomenology, articulated by Husserl around the conception of intentionality; there is the parallelism between natural, or empirical language and transcendental discourse, discussed by Eugen Fink, and to be articulated by Derrida through the emergent concept of writing; there is also a parallelism, for Husserl, between a notion of empirical and of transcendental genesis, which leads Derrida to surmise a problem relation between the sense of genesis and the genesis of sense.

Derrida's approach to these parallelisms moves from a direct reading of Husserl's texts, to considering their refraction in literary composition. The critical delimitation, staged in an aphorism, with its beginning and end, marked off by the intervening placement of numbers ('thirteen', 'fourteen', 'fifteen', 'sixteen'), reproduces in outline the Husserlian gesture of bracketing, or *époche*. This sets aside any naturalising assumptions about what there is, naming and suspending the natural attitude, on which those assumptions depend. This, in turn, permits a reassessment of the status of what presents itself. The startling transposition, here, is from a phenomenology concerned with analysis of the regularities, making perception possible, to a concern with the irregularities which writing opens up, within the grids of meaning made available in a natural language. It is this insistence in Derrida's readings on actual inscription which wrecks Husserl's careful construction of a non-platonising idealism. This key transition effected by Derrida on the phenomenology of Husserl, renders that phenomenology unrecognisable, at least to large numbers of Husserlians, for whom phenomenology is a phenomenology of perception and its impressions, not of impressions and their inscriptions. For Derrida, Husserl's originary impressions are already marked by the Freudian impact of an after-impact, and by the pre-inscription of inscription. This coupled to his insistence on reading Husserl backwards from the later work on history, in *The Crisis of European Philosophy and Transcendental Phenomenology*, rather than forwards from the *Logical Investigations* (1900), disorients the more conventional Husserl reception.

A key element in Derrida's responses to Husserl is indicated in this juxtaposition of the notion of inner time consciousness with that of a spatialisation, in the bracketed phrase: ('the phenomenology of internal time consciousness' or space consciousness). The claim is that Husserl's technique of reducing the proliferating emanations of time to the series of impressions, registered in internal time consciousness, must impose a spatialisation on time, by charting it in a topography of the functioning of consciousness. This supposition is in part borne out by Husserl's attempts in the earlier texts on time, HUA 10, to provide diagrams of time, with a structure of back projections of the impressions given at any one moment, steadily receding and fading into the past, as time presses forward. This structure sets out an open horizon, of present impressions receding into a continuous past, which then provides schemata for projecting meaning expectations into the future. These projections, called by Husserl protentions to mark their relation to retention, circumscribe a horizon of potential meanings, which are certainly corrigible, but which nevertheless set meanings out as systematically organised and internally coherent within a single set of coordinates. Diagrams including protentions are to be found in the *Bernauer Manuscripts*, HUA 33.[27] The time envisaged resembles the block time of a four-dimensional space-time, with the three spatial coordinates supplemented by a series of loops constituting a single sequence of before and after moments. However, this single sequence is rapidly overwritten by a series of sequences in parallel, which then start to cross over one another, in the hypothesis of destabilised parallelisms, between empirical consciousness of time, its grounding in a transcendental instance, called absolute consciousness, and the sharing of that grounding between different bearers of empirical consciousness. These then are inscribed within a series of overlapping inheritances of transmitted meaning determinations, which are variously reactivated by the various bearers of consciousness. To the immediacy of retention, maintained without a break in relation to present perceivings and meanings, Husserl contrasts the contents retrieved from the past, through reactivations of memory. These structures, through which the discontinuity of forgetting and memory may be overcome, he calls 'presentifications', (*Vergegen-wärtigungen*), or bringings into presencing (*Gegenwart*). Much is made of the difference between this notion of the *Gegenwart*, held in place by contrast to the past and the future as dimensions of time, and a concept of presence, as a *Präsenz*, as a present tense in conjugations of verbs, derived from the Latin, for 'being' (*esse*), which is 'near' (*prä*). The former marks the coming into a present moment out of the future, with a dwindling away into the past. The latter is an ontological notion, marking the status of entities as the kinds of entity which present themselves in their being.

This notion of presence is subjected to scrutiny and disruption in Heidegger's analyses of a 'nearness', of this '*prä*' of '*präsenz*', with respect to the supposed distance of ontological determinacy, which conceals that this determinacy is nearer than what presents itself as near: 'Ontically of course Dasein is not only close to us – even that which is closest: we are it, each of us, we ourselves. In spite of this, or rather for just this reason, it is ontologically that which is farthest' (H: SZ 15, MR p. 36). There is then also a dispute about whether empirical entities, pianos or bicycles, are more present than abstract entities, tunes or speed. For Heidegger, however, what is most mysterious, but what is closest of all, is the fact of there being anything at all. For Heidegger, there is the all important distinction between entities present-at-hand, or ready-to-hand, and the mode of existing, Dasein, which by virtue of its existence encounters entities in modes of givenness. Derrida, by contrast, remains unconvinced that the ontological privilege to what is closest is in fact separable from the empirical fact that what is given, is given only in a present empirically given moment. He thus supposes that Heidegger cannot break free from the metaphysics of presence to which he, Heidegger, draws attention. Levinas converts the correlative notion of distance, via an emphasis on asymmetry, into those of dissymmetry and an absolute height, from which the intimation of the divine arrives. This distance does not convert into a notion of proximity.

In *Speech and Phenomenon*, Derrida transposes this diagnosis concerning the reinstallation of a dogmatics of presence to include Husserl, as well, on the basis that intuitions provide immediately given data. However, for Husserl there can be empirically given moments, as *Gegenwart*, only on the basis of there being a transcendentally organised temporal series, held in place as living presence, the notion of a *lebendige Gegenwart*. This, then, provides a fourth meaning for the term 'presence', alongside those of past, present and future, of the grammatical tense, and Heidegger's notions of nearness and distance. This notion of a living present is no longer the punctual 'now' of a time series, and provides a distended notion of time, which is not to be re-subordinated to a notion of absolute duration, given in that notion of the eternity attaching to divinity. However, it puts considerable strain on the notion of 'living', and leads to Derrida's analyses of an immobilisation of life, in the thought of transcendental life, as equivalent to that of transcendental death. His dispute with Husserl concerns the status of both distinctions, life and death, and empirical and transcendental duration, and the distinction between them. The distinction between the temporal 'now' of presence, as *Gegenwart*, and the ontological 'now' of *Präsenz*, is basic to the distinction, key for Husserl's phenomenology, between presentation, in continuous time, and presentification, which overcomes breaks in attention. The latter, in turn, depends on a distinction between continuous memory, called by Husserl retention, and a remembering, which overcomes a phase of forgetting.

In the latter, a meaning content, which has dropped out of view, is brought back into the sphere of a horizon, which delimits what can be attended to. Derrida's notion of spatialisation challenges the status of this horizonality, highlighting how the differences between the elements so carefully distinguished by Husserl are undermined by the presumption that they can all the same be brought back into the sphere of that single horizon. The unity of this horizon cannot be derived from an evidential givenness, but must be assumed as an 'Idea in the Kantian sense'. Derrida seeks out resources in Husserl's writings for a less simple, but perhaps more vivid delineation of the workings of time, as both available and unavailable for interrogation, as itself interrupting the smooth transitions which Husserl hopes to show are equivalent to the passage of time. Derrida thus reads Husserl's account of time, the smooth continuous absolute flow, as covering over another account of time, also in evidence in his analyses, as interruption, discontinuity and indeed syncopation. Derrida develops modes of writing to underline these features of time, as opposed to the smooth flow of a continuous classical prose style, which mimes the continuities of classical accounts of time. These various experiments with modes of writing and composition are in evidence in the writings of the 1980s, in *Glas: What Remains of Absolute Knowledge* (1974), *Truth in Painting* (1978) and *The Post Card: From Socrates to Freud and Beyond* (1980) These connect to the developing mode of writing in the margins which culminates in the invention of *circonfession*, in *Jacques Derrida* (1991) by Geoffrey Bennington and Jacques Derrida. This *circonfession* mimes the movement of Husserl's *époche*, by infinitely multiplying it. The *époche* provides a supposedly neutral framing, available to any reader, through which what there is may be revealed as what there is. The practice of *circonfession*, by example, reveals a framing through which any reader responds to reading, each time uniquely.

The technique of bracketing, Husserl's *époche*, suspends involvement in a world and in activities in that world, in order to reveal its constitutive features, and to reveal how the world is constituted. The former, revealing its constitutive features, is the task of static phenomenology; the latter, revealing them in process of constitution, is the task of genetic phenomenology. The *époche* permits a space of enquiry to be opened, within which the *eidos*, or essence, informing the flow of phenomena, and constraining the possible forms of phenomena, may be revealed. The transcendental conditions for these non-naturalising activities of both bracketing and reduction can then be shown by Husserl to be aspects of transcendental subjectivity. This transcendental subjectivity and its distinctive capacities have a transcendental genesis which takes place in a sphere of transcendentally reduced time. There is thus a contrast to be marked between empirical genesis, which takes place in natural time, and transcendental genesis, which is to be placed within the sphere of a reduced time. The structure indicated by reference to the 'death sentence', remarked in the second aphorism, provides guidance here. For it both declares the necessity of death and, by putting it into language, mimics the gesture of phenomenological bracketing, suspending actuality while appearing to permit that actuality of death to be grasped conceptually. Thus, while appearing to be neutral with respect to what is described, description, on this account, necessarily suspends the flow of processes in order to describe their structure. These moves are opened out for exceptionally close interrogation in the writings of Maurice Blanchot, which in turn have their effects on the writings of Jacques Derrida.

Derrida points to an opposition between flow and structure in his juxtaposition of the notions 'genesis' and 'structure', as terms through which to engage with Husserl's writings. This juxtaposition conflates the more usual pairings, genesis with *eidos*, and history with structure; the crossing of the pairings replicating the movement set up by supposing there to be a parallelism between the result of empirical and transcendental analysis, while supposing that empirical genesis grounds in a transcendental condition. The differences between genesis and structure are made stronger by linking them to a further difference between giving a structural and a genetic account of the emergence of such distinctions. In *The Problem of Genesis in Husserl's Philosophy*, Derrida gives a structural account of genesis, with a consequent foreclosure of the scope of the emergent concept of genesis. Derrida's readings there of Husserl's writings show how they repeatedly present variations of a single aporetic structure, attempting to do justice to the movement of genesis through the inadequate conceptual resources of eidetic phenomenology. By contrast, the reading of the *Introduction to Husserl's 'Origin of*

Geometry' (1962), rather follows the flow and development of Husserl's thought, tracing its genesis, both actual and transcendental. The first reading emphasises the unity of Husserl's thought, identifying a series of stages of his writings, as each time marked by a single repeating *aporia*, concerning the relation between genesis and *eidos*. The second reconstructs stages of Husserl's analyses, to give a context for reading the late essay 'On the Origin of Geometry', written in the course of preparing his posthumous text, *The Crisis of the European Sciences and Transcendental Philosophy*. It also concerns itself with Husserl's borrowing and critiques of other strands of enquiry, in the history of philosophy, principally those of Kant and of Hegel. Derrida's reading in the *Introduction* begins by setting up a distance between Husserl and Kant, but ends by marking a distance between Husserl and Hegel, on how to understand the notion of the absolute. This latter move is summed up in the puzzling phrase, 'The Absolute is passage' (D: 1962, HOG p. 149), to which I shall return. The passage in question is that from temporally circumscribed, empirically given evidence to temporally unrestricted conceptual content, and the problem is to secure its legitimacy. The contrast between structure and genesis is further emphasised by the parallel differences between the notions of static and genetic phenomenology, and between an essence of time, and a genesis of understandings of time, as it unfolds in empirical consciousness.

The time of empirical consciousness is folded into the natural time of days, nights, and appears to have the form of a linear continuity; transcendental time should result from bracketing naturalising assumptions about time, to reveal its essence, by bracketing the natural attitude and then performing the required sequence of phenomenological, transcendental and eidetic reductions. Whether death, chance, or indeed time lend themselves to such treatment, of course, is what remains to be determined. There is also a shift here from analysing Husserl's concept of genesis, to surveying the implications of the emergence of a concept of transcendental historicity, which, while quite distinct from empirical history, is shown in some sense all the same to presuppose an empirical history. Husserl hopes to solve the problems raised by Derrida in relation to empirical and transcendental genesis and, implicitly, in relation to empirical and transcendental time, by invoking this notion of transcendental history. In the next part of my enquiries I shall discuss Derrida's reading of Husserl as suggesting that Husserl undermines the descriptive status of his specific mode of transcendental enquiry by making use of the normative concept of the Idea in the Kantian sense. The question is whether it is possible to keep the two forms of transcendental enquiry separate.

This part of my enquiries then has rehearsed in outline some key contested Husserlian concepts: the historical *apriori*, horizon, intentionality, transcendental constitution, protention and retention; and it has rehearsed, again in outline, some of Derrida's objections to Husserl's enquiries. This provides a background against which to look once again, more closely, at what unites and what divides Derrida's three early readings of Husserl. My main suggestion so far is that Derrida's *différance* articulates a mode of temporality, in which Husserl's requirement of a rewriting of Kant's transcendental aesthetic is in part realised. This rewriting takes up Heidegger's objection to the interminable delay in Husserl's phenomenology of arriving at ontological specification. My second suggestion, even more controversial, is that Derrida makes an inestimable contribution to Husserl's phenomenology by developing a challenge to naturalised notions of the horizon. The horizon is no longer to be understood as a line between earth and sky, as it appears to remain even with Heidegger. It is to be rather understood as a virtual and prosthetic limitation, traced out by the ellipses of meaning and the detours of thought, in the intending and fulfilling of meanings, as held in place in that obscure medium called *écriture*. A distinction between attending to topography as writing, as opposed to topology as pure form, will return for attention in the section thereafter.

Part II

Interrupting Husserl

2.1 The sense of genesis and the genesis of sense

> In Husserl, it is always a question of a difference between the fullness and emptiness of an intuition of meaning, between a more and a less in the plenitude of intuitive presence, in what Husserl calls (a strange figure which would pose so many problems!) the fulfilment (*Erfüllung*) of the intuition. One could translate *Erfüllung* by 'accomplishment', execution, realization or even performance.
>
> (D: 2000, EC p. 291)[1]

The conditions for a fulfilment of the meaning of time and the conditions for a fulfilment of the meaning of genesis are distinct, and it is the latter which is the focus for Derrida's early text *The Problem of Genesis in Husserl's Phenomenology* (1953/54/1990).[2] This citation, from a much later essay published in 2000, draws attention to the bivalence of Husserl's questioning of intuitions of meaning: they are either full or empty, and their protentions may be either fulfilled or disappointed. I shall explore here the manner in which Derrida develops Husserl's own notion of the Idea in the Kantian sense, to reveal an operation of a further Kantian concept, that of transcendental illusion, in which empty intuitions are taken to be fulfilled, or an intuition, fulfilled at the empirical level, is falsely taken to be fulfilled at the transcendental level as well. At times, Derrida seems to suggest that these illusions are necessary for the functioning of meaning and reason. Thus, an analysis of various distinct deployments of the Idea in the Kantian sense, by both Husserl and Derrida, and indeed by Kant, at times covers over an even more telling deployment by Derrida of the Kantian notion of transcendental illusion, as analysed in the Dialectic of the *Critique of Pure Reason.*[3] This refers back to the amphibology of concepts, as discussed at the conclusion of the Analytic of Principles of Concepts, analysing the twin errors of taking an empirical deployment of a concept for a transcendental one, and of a transcendental deployment for an empirical one, errors which Kant ascribes to Leibniz and to Locke respectively (Kant: A 271, B 327). In the first part of this section, I shall explore how Derrida's attention to Husserl's use of the Idea in the Kantian sense provides the outline of an imperfect formalisation for his reading of the strengths and lacunae of Husserl's enquiries, and in the second part of this section, I go on to discuss the workings of this concept of illusion.

The 'sense of genesis' is not a phrase much in evidence in Husserl's writings, but the term 'genesis' is of increasing importance from 1919 onwards. It is introduced in the preparation of the manuscript, *Experience and Judgment: Towards a Genealogy for Logic* (1938), and discussed in the 1921 *Lectures on Transcendental Logic*, translated by Anthony Steinbock as *Analyses Concerning Passive and Active Synthesis* (HUE 9, 2001).[4] However, Derrida surmises that already in the early mathematical writings Husserl is concerned, at least implicitly, with a concept of empirical genesis, in the emergence of the possibility of formulating logical and mathematical concepts. The structure of Derrida's enquiry traces out how in the Prolegomena to the *Logical Investigations*, any suggestion that meaning can be determined by the actual processes through which such concepts are thought is to be rejected in favour of an idea of pure logic.[5] This is the famous critique of psychologism, with respect both to the thinking of mathematics and by extension to the thinking of meaning, more generally. The possibility of actual meanings being governed by a pure grammar, founded in this notion of logic, is what is then to be demonstrated in the course of the following six investigations. Derrida's text discusses these two stages of Husserl's thought in the first two sections of his study, as presenting shifting attitudes to the notion of genesis. The study then goes on in the third section to explore how the notion of genesis, as an empirical concept, is to be replaced by a notion of genesis as a transcendental concept. The fourth section explores how Husserl's notion of history is driven by the requirement for a reactivation of an initial genesis of meaning. However a unity for these various notions of genesis is as difficult to set out as is a unity on the deployment of the notion of the 'Idea in the Kantian sense'.

In fulfilling its sense, the status of genesis as genesis is apparently eroded, for it ceases to trace the emergence of meaning and becomes a meaning, as fully determined. If the sense of genesis is taken to be the genesis of sense, the movement of chiasm intervenes here to stall the affirmation of the Husserlian procedure, and displaces any securing of its argument into an account of how sense is fulfilled in history. In the introduction to *The Problem of Genesis*, Derrida declares: 'The genesis of sense is always *apriori* converted into a sense of genesis that supposes a whole history of philosophy' (D: 1954, PG p. 3). This is at first sight a surprising claim, for it might seem that it is precisely the genesis of sense which marks out transcendental history, if transcendental history consists in those moments at which thought contents come available for the first time, changing the articulation of possible thought and changing the manner in which world constitution takes place, in part or indeed as whole. There are two crucial shifts here. The first is that between thinking of genesis, as a genesis of a specific thought content, for example the notion of number, to taking the notion of genesis as the thought of there being meaning contents at all. The second shifts thinking of the first historical occasion of a thought content from being thought as an empirical accident, to thinking of it as a transcendental necessity. This introduces the difference between empirical and transcendental history. The stalling of Husserl's analysis occurs if the sense to be reactivated is taken to be that of the second notion of genesis, and there may be here in operation an antinomy of reason, whereby partial sense can be thought, in relation to a concept of transcendental genesis, but not the concept of sense itself. At this stage, it seems that Derrida is construing the Husserlian conception here as a single thought of meaning, science and truth, for which reactivation is required, rather than as a series of conceptions, which may not be synthesised into a single totality.

Derrida returns to the relation between sense, genesis and the process of history later in the third part of his text, where he surmises:

> Every history announcing itself is reduced *apriori* to its phenomenological and intentional sense, to a sense which it did not create in its authentic genesis but which pre-exists it, envelops it, and continually informs it. Its possibility is the modification of an originary constituting activity. The eidetic rigor is saved in this way, but it is by altering or suppressing genesis.
>
> (D: 1954, PG pp. 143–44)

He concludes the chapter by writing:

> After *Cartesian Meditations*, where it makes its first appearance, philosophical teleology will occupy a privileged place in Husserl's thought. In the important cycle of the *Krisis* and in the '*Origin of Geometry*', it develops into a veritable philosophy of history. It is this which we must examine to conclude, asking ourselves to what degree this philosophy of history, bringing to a close the system of transcendental phenomenology, at the same time and at the same moment sanctions the unsurpassable depth and the irreducible insufficiency of Husserl's philosophy of genesis.
>
> (D: 1954, PG p. 149)

This supposed smooth transition from teleology to a philosophy of history is remarkable, for it presumes that a teleology of fulfilments of meaning intentions has the same temporality as the unfolding of a meaning in a directedness of history. It is this Hegelian merging of two distinct processes, one analysed by Husserl, and the other analysed by Kant, which Derrida's reading of Husserl brings into question. For while there are no doubt places where Husserl seems to endorse the conflation of the two, they are in fact distinct for him, since the fulfilment of meanings takes place in transcendental time, while the historical development takes place in natural time.

The key suggestion in *The Problem of Genesis* is that the attempt to articulate genesis is thwarted at each turn by the imposition on the phenomena of genesis, of the results of the operation of reduction, which in each case according to Derrida, reveals a sense given *apriori*, in advance of its genesis. The transition from *The Problem of Genesis* to the *Introduction to Husserl's 'Origin of Geometry'* is a transition from an analysis of Husserl's accounts of genesis to a discussion of Husserl on history. Marking differences between empirical and transcendental genesis, as grounded in a distinction between empirical and transcendental history, appears to solve this problem. However, the problem returns once the stability of this further distinction between genesis and history is shaken, in the challenge developed in *Speech and Phenomenon* to the distinction between empirical and transcendental meaning. Thus the *aporia* set out in *The Problem of Genesis* appears to be resolved by insisting on a distinction between empirical and transcendental genesis, as supported by the notion of transcendental history, but considerations introduced in those two subsequent texts also undermine the stability of that distinction. Later in this study I shall show how Derrida provides the elements of a resolution of the *aporia* he identifies here in Husserl's text, in a transcription of the distinction between phenomenological psychology and transcendental phenomenology, into one between empirical and transcendental experience, an experience within time and an experience as endurance, constituting time. Here, what is in question is the hesitancy and precipitation with which he reads Husserl, first setting up a critique, then appearing to provide a defence against that critique, and then in a third text apparently definitively rejecting the Husserlian programme, but only in order to reinstall some of Husserl's central concerns. This third text of course is *Speech and Phenomenon: Introduction to the Problem of Signs in Husserl's Phenomenology* (1967).[6]

The remark given as a citation at the head of this section, concerning fullness and emptiness, invokes the Husserl of the *Logical Investigations* and the distinction drawn there between sensory fulfilment of immediately given empirical contents, and categorial intuition of eidetic contents. The meanings of so-called syncategorematic terms, such as 'and' and 'but' are more like the latter than the former. They acquire meaning from their role in the articulation of structured meanings, in sentences, as opposed to the meanings attaching to supposedly discrete concepts. The affirmation in the sixth of the *Logical Investigations* of this notion of categorial intuition permits Husserl to consider the articulations of meaning in propositions to be the basic unit of meanings, as already intimated in the transition from Investigation Three, on wholes and parts, to Investigation Four, on dependent and independent meanings, and to the notion of logical grammar. Thus neither Husserl, nor indeed Derrida, when reading Husserl, expects there to be a meaning in the form of a single simple diagram, or image for the term 'genesis'. Nevertheless, the question of how to conceive of fulfilling an intuition of its sense remains moot, as does the shift from constituting meaning in sentences, committed to articulation in discrete distinct natural languages, and constituting meanings in propositions, in which meaning is no longer inextricably linked to any one such natural language.

Once the question is rephrased as: what are the modes of givenness of genesis, the question becomes more pointed. For if temporality is the basic form of modes of givenness, then the modes of givenness of genesis, itself a mode of temporality, must be some further mode of temporality. This doubling up of temporal determinations is pursued by Husserl through the labyrinthine analyses of the *Bernauer Manuscripts* (HUA 33).[7] While transcendental phenomenology remains descriptive, there is a shift from ascribing to what is described an empirical evidence, to ascribing to it an eidetic or categorial evidence. Derrida pushes this move one step further, in *The Problem of Genesis*, and attends to the modes of givenness of Husserl's own theoretical shifts, and, especially, to the manner in which the notion of genesis remains continually at work, in the various stages of Husserl's thinking, although in different guises. In the preface written in 1990, Derrida diagnoses 'the unity of his search' as resulting from the relation between empirical and transcendental genesis, the former prior in the natural time of the development of Husserl's writings, the latter prior in the order of concepts. In the original preface he writes:

> In a word, if the theme of transcendental genesis appeared at a certain moment in order to understand and found the theme of empirical genesis that preceded it in natural time, we need to ask ourselves about the meaning of this evolution.
>
> (D: 1990, PG p. xix)

The parallelism between these two acquires its fullest articulation only once the full context of Derrida's explorations of the implications of the other two parallelisms, between empirical and transcendental language, and between empirical psychology and transcendental phenomenology, has been set out.

This unity of the enquiry is described by Derrida, in the 1990 preface, as 'a sort of law', which

> will not have stopped commanding everything I have tried to prove, as if a sort of idiosyncrasy was already negotiating in its own way a necessity that would always overtake it and that would have to be interminably re-appropriated. What necessity? It is always a question of an originary complication of the origin, of an initial contamination of the simple, or an inaugural divergence that no analysis could *present, make present* in its phenomenon or reduce to the point-like nature of the element, instantaneous and identical to itself.
>
> (D: 1990, PG p. xv)

This originary complication of the origin is what Derrida discovers in Husserl. Even the omni-temporal meanings of exact eidetic essences turn out for Derrida to have origins which arrive out of the future, in the complication of originating meaning intuition by its multiple future fulfilments. The shift from complication, to contamination, to an invocation of 'an inaugural divergence', or *clinamen*, stages a series of shifts in the manner in which this trajectory is to be thought, resulting in a rewriting of necessity itself. This inaugural divergence is thought by Derrida in relation to the responses to Epicurean atomism provided by Freud and by Nietzsche. The first staging of this is to suppose that Derrida both does and does not seek to inscribe his enquiries within the framework set out by Husserl for the development of phenomenology.

The three texts on Husserl composed by Derrida between 1953 and 1967 have quite distinct forms, which map on to distinctions in the manner in which the Idea in the Kantian sense is deployed in them. The first provides an interpretative overview; the second promises a detailed textual commentary; and the third, *Speech and Phenomenon*, as remarked, declares itself to be neither interpretation nor commentary. The interpretative overview mobilises a conceptual structure, the Idea in the Kantian sense, in part borrowed from the texts themselves, to impose and justify a unity on its readings. The commentary on the essay 'On the Origin of Geometry' displays the virtues and vices of the genre, of seeking to reconstruct a context for and to follow the line of thought of the other. It thus remains the more constrained, remaining within the conceptual limits attributable to the commented text, but it opens up a distance between Husserl's use of the term and that of Kant. *Speech and Phenomenon* by contrast has the form of a condensation of Husserl's writings, resulting famously in a distortion of the concept of the living present. This distortion is perceived as a violent reading, but the question to be posed concerns the degree and effects of this violence: for violence in itself cannot be taken to be necessarily an objection to the reading. It is worth remarking the differences of their distinct forms, because they prevent any easy following through of a single set of thematics from one text to the next. The ambition of each text is markedly different, serving as correctives each to the one before.

The first, *The Problem of Genesis in Husserl's Philosophy* as remarked identifies a structure of a repeating *aporia,* which prevents Husserl's texts from advancing to a conclusion satisfactory to their author. The second text explores the strain imposed on some of the central concepts of phenomenology by the attempt to describe how geometrical innovation has been possible. In the third, Derrida announces the double movement away from, and back towards metaphysical commitment, which he finds in Husserl's texts. In the second chapter of this third text responding to Husserl, entitled 'The Sign and Signs', he remarks:

> The historic destiny of phenomenology seems in any case to be contained in these two motifs: on the one hand, phenomenology is the reduction of naïve ontology, the return to an active constitution of sense and value, to the activity of a *life* which produces truth and value in general through its signs. But at the same time, without being simply juxtaposed to this move, another factor will necessarily confirm the classical metaphysics of presence and indicate the adherence of phenomenology to classical ontology.
>
> (D: 1967, SP pp. 25–26)[8]

The possibility that the sign here is not just that of Saussure, but also that of Nietzsche, the sign permitting a diagnosis of what is distinctive of an epoch, marking the arrival of the new, deserves to be marked up. This 'classical metaphysics of presence' is the phrase indicating the importance for Derrida of disputing the differences between the notions of time at work in the operations of presentation, and of presentification, and between the present as present moment, and the present as dimension of time, contrasting to past and future. The claim by Husserl that these various aspects are unified in a thinking of the unity of the living present for Derrida rather confirms than disproves his diagnosis, for

Derrida supposes that the conception of the living present exceeds all possible intuitive content with which the meaning intuitions might be fulfilled. The status of this key Husserlian concept is thus Kantian, that of an Idea in the Kantian and not in the Husserlian sense.

The later text, *Le toucher, Jean-Luc Nancy* (2000) sets out yet another strategy of reading, by emphasising a notion of tangents, whereby the readings touch on Nancy's concerns, and thereby reveal how both that text and Nancy's texts touch on and seek to display an impossible delimitation of both phenomenology and philosophy. It is thus dedicated to a response to phenomenology, via the reading of the writings of Jean-Luc Nancy, but it opens by invoking Aristotle in book 2 of *De Anima*, on the puzzle (*aporia*) of whether there are many senses of touch, or just the one (422b).[9] For Aristotle, the lived body is the medium and not the organ of touch, unlike the relation between the eye and vision. Derrida discusses Nancy's thinking of a limit case of touch, or sensory contact, which touches an untouchable limit: a touching which does not touch.[10] The first half, the reading of Nancy, reveals a continuity between Nancy's concerns and the analyses of Husserl, as presented in *Ideas Two*, of the inorganic, the organic and the specifically embodied status of human lived experience, which prompted some of Merleau-Ponty's reflections on embodiment. This first half is then supplemented by a series of discussions, in the second half, under the title 'Tangents', of other phenomenological writings. The delimitation achieved is of a specifically French reception of phenomenology, through a series of references to French studies in phenomenology. This demonstrates that once the practices of footnoting set in, there is necessarily a restriction of the scope of enquiry, setting out specific concerns and questions. These tangentially related enquiries concern the distinctive commitments of the principal French responses to and reconstructions of Husserl's phenomenology, those of Levinas, of Merleau-Ponty and especially of Didier Franck, in his study *Chair et corps: sur la phénoménologie de Husserl*, or *Flesh and Body: On the Phenomenology of Husserl* (1981).

This last explicitly poses the problem of tracing shifts in meaning resulting from transposing Husserl's distinction between the quality, distinctive of lived bodies, *Leiblichkeit*, and that of bodily extension, *Körperlichkeit*, into the notions of *chair*, flesh, and *corps*, body. The difficulty of even marking this distinction in English should be recalled, and the manner in which for the French, but not for the English, there is an immediate connection back to discussions of Descartes on a mind/body dualism is worth considering. The shift from the first half of the text on Nancy, to the second half, transposes the direction of discussion from that of seeking from within a reading of Nancy and of Husserl's texts, to escape the apparent paradox of conceptions of transcendental life miming a death in life, to that of seeking from a stance, apparently beyond the problems of Husserl's phenomenology, to reveal that those problems are still the terms of reference for phenomenology, and more generally for philosophy. Derrida marks up a specifically French concern with a marked contrast between an order of Cartesian ideas and an order of mechanically given material instantiation, which readings of Husserl at first challenged, and then appeared to reinstall, through the workings of a contrast between absolute transcendental life, and absolute transcendental death. The machine is figured as the typewriter of 'Typewriter Ribbon: Limited Ink. (2)' and will be discussed in Part V of these enquiries. The discussion of Nancy is constructed with a series of responses to the writings of Nancy, as it were surrounded by the tangentially arranged responses to Husserl, by this series of leading French phenomenologists, writing in a context framed by a distinctively French reception of Descartes. This elaborate, if inconclusive, textual structure draws attention to the manner in which Derrida's three early texts on Husserl, when juxtaposed, have a similarly unsettling effect on the supposition of there being a single point of vantage from which the reading of Husserl might take place. There is thus a line of connection from the practice of *circonfession*, writing around the texts of Bennington and Augustine, and this mode of tangential writing, around the texts of Nancy, and the writings constituting the specifically French reception of Husserl, through Levinas and Merleau-Ponty, Didier Franck and Jean-Luc Nancy.

In each of these earlier texts, there is a massive work of distillation, condensation and displacement imposed on the proliferating texts through which Husserl sought over and again to arrive at an exposition of his philosophical programme, while also always starting again from the beginning. I shall begin the exposition of the three earlier texts by setting out the distinct ways in which in each Derrida invokes the Idea in the Kantian sense. The central problem is that of delimiting domains of enquiry, and in the first text, the Idea in the Kantian sense is deployed in relation to four distinct conceptions of infinity, and of non-finitude, to trace out a delimitation of four distinct conceptual horizons, for the supposed recurrent *aporia* with respect to Husserl's analyses of genesis. For the second text, the deployment of the Idea in the Kantian sense provides a means of opening up a gap between Husserl's enquiries and those of Kant, while offering a different series of determinations of that Idea, from that offered in the former. It is deployed to bring into focus various different senses in the notion of infinitisation, which is, for Husserl, the innovation distinctive of rigorous scientificity. For the third, its deployment provides a means of bringing together the elements of a critique of Husserl's phenomenology, to make clear the occurrence of an alterity not controlled for by Husserl, in the dependence of the distinctively Husserlian concept of ideality on the Kantian Idea, which cannot be phenomenologically grounded. This opening on to a theoretical heteronomy is then underlined by the appeal to Freud, midway through that text, and by appeals to the Nietzschean, Bergsonian and Heideggerian critiques of metaphysics, appended as a footnote to the description given above of a double movement of Husserl's phenomenology.

The shift from the overview of *The Problem of Genesis*, to the commentary at close proximity to the text, on Husserl's late essay 'On the Origin of Geometry' also marks a transition from an invocation of a more strictly understood conception of the Idea in the Kantian sense, closer to its deployment by Kant himself, to the apparently more diverse series of invocations of it, made by Husserl himself in sections 74, 83 and 143 of *Ideas One*.[11] The shift from interpretative overview to commentary also introduces a difference in the temporal mode which structures the writing. The first text, while concentrating on the question of genesis does not problematise the genesis of its own meaning, and thus minimises any constitutive role for time or genesis in the writing of the text. In the commentary on 'On the Origin of Geometry', there is more attention to the time of writing, with the commentary paradoxically extending to four times the length of the commented text. The time of writing is made even more salient in the third text, which enacts a very marked deceleration and acceleration of pace of reading, with the first four chapters dedicated to a reading of the first of Husserl's *Logical Investigations*, 'On Some Essential Distinctions', and the remainder speeding through to draw out a general problem about the thinking of time, in Husserl's phenomenology as a whole, arriving at the claim about the repression, within that thinking, of otherness. In its invention of the figure 'the supplement of the origin', however, this third text makes an implicit claim to be retrieving and reactivating a hidden truth of the originary Husserlian insight into how meaning is possible, how enquiry is cumulative across generations and how thought can become aware of itself. This notion of a necessary 'supplement of the origin' reveals the impact of a reception of Rousseau's enquiries on a reception of Husserl.[12] However, this text, which is the most objectionable to Husserlians, oddly, retains more of a commitment to affirming the aim of Husserl's enquiries, affirming a task for philosophy, if criticising the terms of his analysis.

In these texts Derrida explores, first from within Husserl's enquiries, and then with an increasing distance from them, the problem of invention and the possible arrival of new thought contents. This is explored by Husserl in the discussion of geometry, through the concept of *Erstmaligkeit*: that something can be genuinely thought for the first time. The inventions of geometry are either more like the discoveries of cosmology, where a star or planet is found, which has always been there, or, like the invention of the language of James Joyce's *Finnegans Wake*, a potentiality, which is revealed as potentiality only once it has been activated. Theological re-appropriations of Husserl would argue that geometry for him is on the model of the former. The more radical reading of religion, as always inventing a relation to the divine for the first time, permits his thinking of geometry to have the latter form. In the middle of his *Introduction to Husserl's 'Origin of Geometry'* Derrida suggests but does not pursue the question of the relation between literary and geometrical invention, by contrasting the insistence on the univocal with respect to meaning and language in Husserl's text, and on the equivocal with respect to language in that of Joyce.[13] In section 10 he also begins a discussion of the significance of Husserl's invocations of a concept of God, to which he returns in the discussion of the thought of Jean-Luc Nancy and of Jean-Luc Marion, in the 1980s and 1990s.

More salient here is the question, whether the stabilisation of meaning in ideality is a necessary fiction, conforming to the logic of an ideality of a literary invention, rather than to the model of geometrical essences. The role of the Idea in the Kantian sense in providing non-phenomenological support for this key notion of Husserl's phenomenology gives support to this hypothesis. It is invoked thus towards the end of the last chapter of *Speech and Phenomenon*, chapter 7, 'The supplement of the origin':

> That Husserl always thought of infinity as an Idea in the Kantian sense, as the indefiniteness of an '*ad infinitum*', leads one to believe that he never *derived* difference from the fullness of a *parousia*, from the full presence of a positive infinite, that he never believed in the accomplishment of an 'absolute knowledge', as the self-adjacent presence of an infinite concept in *Logos*. What he shows us of the movement of temporalization leaves no room for doubt on this subject: although he had not made a theme of 'articulation', of the 'diacritical' work of difference in the constitution of sense and signs, he at bottom recognized its necessity. And yet, the whole phenomenological discourse is, we have sufficiently seen, caught up within the schema of a metaphysics of presence which relentlessly exhausts itself in trying to make difference derivative.
>
> (D: 1967, SP p. 101)

Derrida thus ties together the themes of the Idea in the Kantian sense, Husserl on infinity and temporalisation, and the diagnosis of a recurrence within Husserl's phenomenology of a 'metaphysics of presence'. This builds up towards the claim '*la différance infinie est finie*', roughly translatable as 'infinite *différance* arrives in finite contexts, circumscribed by the conjunction of texts, in which it is to be traced'. To be thought at all, the thought is temporally stabilised by a certain privileging of one set of texts and modes of enquiry, over others.

On the preceding page, Derrida marks a connection more explicitly from Husserl's deployment of the Idea in the Kantian sense to the notion of ideality and its deferred fulfilment. There is a shift from writing of infinity as 'an Idea in the Kantian sense' to writing of an ideality of meaning as having the 'form of the Idea in the Kantian sense', a shift not so obvious in the English translation, which I shall therefore marginally amend. For it points to a difference between Husserl's use of the Idea in the Kantian sense, and Kant's own deployment of it. There is a shift from invoking the Idea in the Kantian sense, as the concept deployed by Kant, to thinking of it as a concept internal to Husserl's philosophy, and a further shift, reconstructing the movement in question, in the invention of the figure, the supplement of the origin, introduced in the last section of *Speech and Phenomenon*: Derrida opens up this invocation of the Idea in the Kantian sense in *Speech and Phenomenon*, by remarking the pivotal role of the notion of the living present, for the articulation of the notion of ideality:

> We have experienced the systematic interdependence of the concepts of sense, ideality, objectivity, truth, intuition, perception, expression. This common matrix is being as *presence*: the absolute proximity of self-identity, the being-in-front of the object available for repetition, the maintenance of the temporal present, whose ideal form is the self-presence of transcendental *life*, whose ideal identity allows *idealiter* of infinite repetition. The living pre-

> sent, a concept that cannot be broken down into a subject and an attribute, is thus the conceptual foundation of phenomenology as metaphysics.
>
> (D: 1967, SP p. 99)

There is, however, a disruption of this living present: 'While everything that is *purely* thought in this concept is thereby determined as ideality, the living present is nevertheless in fact really, effectively etc. deferred *ad infinitum*. This *différance* is the difference between the ideal and the non-ideal' (D: 1967, SP p. 99). And he continues on the next page: 'As the ideal is always thought by Husserl in the form of the (trans. mod.) Idea in the Kantian sense, this substitution of ideality for non-ideality (trans. mod.), of objectivity for non-objectivity is infinitely *deferred*' (D: 1967, SP p. 100).

The questions Derrida supposes Husserl to be unable to answer are the following: 'How can we conceive this difference? What does *ad infinitum* mean here? What does presence mean, taken as *différance ad infinitum*? What does the life of the living present mean as *différance ad infinitum*?' (SP p. 101), and in the formulation he makes use of his own neologism to underline a conversion of contingent temporal delay into necessary temporal postponement, of any supposed fulfilment of meaning. This result links back to the first deployment of the term, in the opening pages of the introduction to *Speech and Phenomenon*, where Husserl's use of the Idea in the Kantian sense is specified in terms of a teleological function. This displaces the time of the supposed fulfilment of the meaning intuition of an originary evidence, from a past occasion, to be reactivated in a present moment, into a future possibility, to be aimed for and anticipated in present use. The question then arrives: what distinguishes this surmised past moment from the surmised future moment, if not a presumption that there is a reprise of a natural ordering of time, which alone can keep past and future distinct? Is there a temporal difference underpinning this shifting of the determination of temporal order, from a retrieval from a given past of a determinate content into a present moment, into the anticipation in the present moment of a possible future completion of determinate content? Or is it the ordering itself, the direction of the movement, which is basic? There is, however, also an ambiguity in the notion of presence, to be marked up:

> The factor of *presence*, the ultimate court of appeal for the whole of this discourse, is itself modified, without being lost, each time there is a question of the presence (in the two related senses, of the proximity of what is set forth as an object of intuition, and the proximity of the temporal present which gives the clear and present intuition of the object its form) of any object whatever to consciousness, in the clear evidence of a fulfilled intuition. Indeed, the element of presence is modified whenever it is a question of self-presence in consciousness – where 'consciousness' means nothing other than the possibility of the self-presence of the present in the living present.
>
> (D: 1967, SP p. 9)

This is important because it grounds the apparent ambiguity between presence as the presence of the object to consciousness, and the presence in time of the present, in a third instance of presence, the self-presence to itself of consciousness, which in turn is to be grounded in the all important, but ambiguous notion of the living present, providing access to the immediate past of retention and the immediate future of protention.

While these moves are made explicit in Husserl's 1904–05 *Lectures on Inner Time Consciousness*, subsequent to the exposition of the *Logical Investigations*, Derrida reads this as a contingent and not a necessary lack of temporal synchrony, such that they may be thought to be already implicit within the earlier text. Similarly, he supposes that the operation of reduction, at least as phenomenological reduction, is already implicitly at work in the *Logical Investigations*. This attribution to earlier texts of distinctions made explicit only in later texts can produce the effect of a violent conflation of texts, and a sense of disorientation, as Derrida's readings leap from one text to another in pursuit of a conceptual structure, stretched out between them. The reduction of natural time reveals a givenness of time itself, in self-consciousness, as the processes of protention and retention, constituting the living present, around the originary impression. Thus the possibility of temporal order, as naturally given ordering, is described by Husserl as depending on these conceptually prior temporalisations of self-consciousness. By calling these processes 'auto-affection', Husserl and Derrida draw attention to a link back to the Kantian transcendental unity of apperception, as guarantor of a unity of the self, given in advance of empirical givenness. However, for Husserl, if not for Derrida, Husserl's analyses constitute an advance over those of Kant, since they seek to demonstrate the constitution of this unity, rather than inferring it as transcendentally given.

What emerges as a problem for Derrida is the presumption of a full coincidence in time with itself in a given present, which he rather supposes to be split between that which is, and that which is not self-coincident, and split again between distinct levels of consciousness, as thematised by Freud, which are not mutually accessible, and therefore do not take place at the same time, or in the same temporal series. Derrida continues:

> Every time this element of presence becomes threatened, Husserl will awaken it, recall it, and bring it back to itself in the form of a *telos* – that is the (trans. mod.) Idea in the Kantian sense. There is no *ideality* without there being an Idea in the Kantian sense at work opening up the possibility of something indefinite, the infinity of a stipulated progression or the infinity of permissible repetitions. This ideality is the very form in which the presence of an object in general may be indefinitely repeated as the *same*.
>
> (D: 1967, SP p. 9)

Thus the Husserlian concept of ideality is supposed to depend on the Kantian insight formalised as 'the Idea the Kantian sense', but there is a shift here, from invoking the Idea in the Kantian sense, as the Idea in the form of a *telos*, a goal to be aimed for, versus a multiplication of ideas, one governing each ideality, versus a third instance of the Idea in the Kantian sense, as governing the concept of an object, in general. The importance of this concept of an object in general then would be that it promises, perhaps misleadingly, to combine all three senses. The question which Derrida might have posed, as he does in *Speech and Phenomenon* to the concept of sign, and in *The Problem of Genesis* to the concept of genesis, is whether or not there is a single Idea here, governing all the deployments of the concept 'Idea in the Kantian sense', or whether the deployment of this term trades on ambiguity. The cumulative effect of his writings on Husserl, is to pose this question to Husserl's use of the term.

In his *Introduction to Husserl's 'Origin of Geometry'*, Derrida explores an ambiguity between the deployment of the Idea in the Kantian sense, with respect to the omni-temporality of eidetic structures, discussed by Husserl in section 74 of *Ideas One*, and its deployment with respect to a trajectory of actual human enquiry, taking place in a history, for which a teleology of reason provides the guide and direction, and which is closer to the deployment of the Idea in the Kantian sense, in section 143 of *Ideas One*. The third deployment of the Idea in the Kantian sense, by Husserl in section 83 of *Ideas One*, concerns the unification of the stream of experiences, which constitutes a unity of consciousness, and, by implication, provides the basis for an account of the continuity of time, since primordial time is taken by Husserl to be constituted in the transcendental consciousness, the unity of which is thus secured. Failure to distinguish between these three distinct deployments, or supposing that they can be shown to be interdependent, brings the reading of Husserl into proximity with the movements of the Hegelian idea. It is thus important to attend carefully to a number of distinctions drawn by Husserl and attended to by Derrida and, as it turns out, by Ricoeur, before Derrida, in order to show how, for Husserl, they remain separate processes.

The relation for Husserl between actual history, that is empirical history, taking place in a natural time series, and its transcendental conditions and transcendental form, as set out in the form of a teleology of reason and in terms of a reduced notion of time, remains in the end under-determined, thus explaining why for example Fink turns to Hegel for an account of how to think the relation between the two. In his *Introduction to Husserl's 'Origin of Geometry'*, Derrida cites Walter Biemel's conclusion, to be found only in the French version of the relevant essay: 'Husserl's essays which try to grasp historicity thematically can be considered as failures' (D: 1962, HOG p. 116). It is, however, more consistent with Husserl's own procedure to suppose that empirical history and transcendental history, like the other various parallelisms between the empirical and the transcendental already remarked on, must be thought of as two ways of thinking about and organising one and the same series of phenomena, the empirical series taken as naturally given, and the same series taken as constituted in the activities of transcendental consciousness. The transcendental consciousness which might be able to constitute such a transcendental history is of course an inter-subjective, inter-generational consciousness, subtending a theoretical possibility which Husserl supposes human life in some strong sense to exemplify, although perhaps granted his invocations of the concept of God, not to exhaust.

Intriguingly at this point Derrida seeks to defend Husserl against the complaint of an inadequacy in the thinking of history:

> If the thematization of the apodictic invariants and of the historical a priori was at fault, would not that be in comparison with *history* rather than with *historicity*? The failure would then be flagrant if, at some moment Husserl was to become interested in something like history. He never seems to have done that. Would not then his original merit be to have described in a properly *transcendental* step (in a sense of that word which Kantianism cannot exhaust) the conditions of possibility for history which were at the same time *concrete*? Concrete, because they are experienced under the form of *horizon*?
>
> (D: 1962, HOG pp. 116–17)

And with this gesture Derrida suggests that Husserl must be taken at his word in the proposal to reduce the temporal horizon of empirical history, as opening into the future, into a horizon concerning 'the totality of possible historical experiences'. Derrida continues:

> Horizon is the always 'already' there of a future which keeps the indetermination of its infinite openness intact (even though this future was *announced* to consciousness). As the structural determination of every material indeterminacy, a horizon is always virtually present in every experience – for it is at once the unity and the incompletion for that experience; the anticipated unity in every incompletion. The notion of horizon converts critical philosophy's state of abstract possibility into the concrete infinite potentiality secretly presupposed therein. The notion of horizon thus makes the apriori and the teleological coincide.
>
> (D: 1962, HOG p. 117)

With this remark, Derrida concludes section 8 of his *Introduction*, leaving for further discussion the emergent distinction between delimiting domains of possible experience, and articulating a singular experience, of a passage to the limit in which such delimitation must consist. Such a distinction would mark a difference between an empirical and a transcendental experience, the one taking place within constituted limits, and the other constituting the limit. In subsequent writings Derrida mobilises the latter to disrupt the former. This distinction between an empirical and a transcendental experience becomes more important for Derrida than any distinction between empirical and transcendental history, or indeed than the distinction between empirical and transcendental genesis, with which his readings of Husserl begin.

In the preceding section, section 7, Derrida has discussed Husserl's deployment of the notion of writing as the medium in which originary meaning intuition can be transmitted for reactivation:

> The possibility of *writing* will assure the absolute traditionalization of the object, its absolute ideal Objectivity – i.e., the purity of its relation to a universal transcendental subjectivity. Writing will do this by emancipating sense from its *actually present* evidence for a real subject and from its present circulation within a determined community. 'The decisive function of written expression, of expression which documents, is that it makes communication possible, without immediate or mediate address: it is, so to speak, communication become virtual' (p. 164, modified).
>
> That *virtuality*, moreover, is an ambiguous value: it simultaneously makes passivity, forgetfulness, and all the phenomena of *crisis* possible.
>
> (D: 1962, HOG p. 87)

Thus meanings, as preserved in writing become separate from actual meaning intendings and fulfilments, in actual empirical consciousness. The meaning intentions and fulfilments at the level of transcendental consciousness are attributed to a virtual sphere of possible fulfilments in actual human experience. In this way, it is possible for the meanings of theoretical formulations to be lost, even when the formulations in script or symbol are preserved. The generalisation of such loss is what Husserl diagnoses as the crisis of European civilisation, which appears to no longer affirm, or even intend the value of its own discoveries, the discovery of the distinctive domains of science, in the infinitisations of human thought contents, beyond the finite compass of human attention spans.

This is taken up by Derrida, in the following way:

> after having presented the capacity of reactivation, Husserl does not fail to ask the serious question of its *finitude*. In a science like geometry, whose potentiality for growth is extraordinary, it is impossible for every geometer, at every instant and every time he resumes his task after necessary interruptions, to perform a total and immediate reactivation of the 'immense chain of foundings back to the original premises' (p. 166, modified). The necessity of those interruptions is a factual one (sleep, professional breaks and so forth), which has no sense compared with geometrical truth but is no less irreducible to it.
>
> (D: 1962, HOG p. 105)

Derrida then poses a series of questions to the coherence of this thought, and adduces: 'But for Husserl, as we know, that finitude can *appear* precisely in its primordiality only given the Idea of an infinite history' (D: 1962, HOG pp. 105–06). Empirical history occurs and can be made sense of only because there is a structure of transcendental history, which opens the finitude of empirical history out on to the non-finite thought contents of human ideals and abstract thought contents. There is then here a double infinitisation: geometrical idealisations permit of an infinitising of the reactivating ability; but the idealisation itself has for its correlate an infinite idea. There are then three incidences of the Idea in the Kantian sense here: it secures the thought from the inadequacy of its formulation in natural language; it secures the thought from the inadequacy of the reactivation of meaning intentions; and its secures the thought from the inadequacy of the thinking, which has up until that point been dedicated to it. It is this last service which is the paradoxical one, for it evokes the temporal structure of the arrival in time of a thought structure which on its arrival turns out to be omnitemporal.

In section 9, Derrida remarks a difference between those works which suppose an already constituted temporality, such as *Ideas One*, and, as he claims, *Experience and Judgment: Towards a Genealogy of Logic*, and those works in which the constitution of temporality is precisely also in question. Derrida distinguishes between empirical history, and a simple eidetic of history, on a parallel with the *eidos* of any other human or natural science, delimiting that which falls within its scope, as opposed to a stronger notion of transcendental historicity which is no longer one science, or regional ontology amongst others, but the dimension within which all constituting activity itself takes place. The tension in this thinking is described thus in the last sentence of Derrida's section 9, to which he then attaches a lengthy footnote:

> That this constituting history may be more profoundly constituted itself, such is, no doubt, one of the most permanent motifs of Husserl's thought; also, one of the most difficult, for it accords badly with that of a historicity which (as Husserl said more and more often) traverses everything through and through, and first of all the *ego* itself.
>
> (D: 1962, HOG p. 121)

The footnote to this reads:

> All these difficulties seem concentrated to us in the sense that Husserl gives to the expression '*transcendental history*', which he utilises (to our knowledge) only once, in an unpublished manuscripts of Group C (C 8 II, October 29, p. 3): thus the question concerns the intermonadic relation (always considered in itself, of course, as an intentional modification of the monad in general in its primordial temporality), a relation thanks to which the constitution of a common world becomes possible. This relation structurally implies the horizon of the history of the spirit, past and future; the latter discovers for us what perception cannot give us.
>
> (D: 1962, HOG, fn 134, p. 121)

These considerations lead up to the astonishing remark which Derrida makes in the last section of his *Introduction*, 'Intentionality is traditionality' (HOG p. 150). For there can be an intending of meaning only on the basis of the transmission of meanings across historical contexts and trans-generationally. This drastic transformation of one of Husserl's basic terms invites a reading of the subsequent *Speech and Phenomenon* as proposing the further rewriting, 'Intentionality is *différance*'. I shall explore this further in the next section of this chapter.

For this section, there remains one more task, of putting in place one further consideration concerning the Idea in the Kantian sense.

In section 10 of his *Introduction*, Derrida invokes Paul Ricoeur's reading of Husserl again, in the course of a discussion of distinct kinds of infinitisation. For Ricoeur there is a shift in Husserl's analyses between what is set out in 'The Vienna Lecture', more or less contemporaneous with the essay 'On the Origin of Geometry' and what is set out in the main text, on which Husserl was working at the time, *The Crisis of the European Sciences and Transcendental Phenomenology*.[14] There is a shift from attributing to the Greeks the inauguration of infinitisation, to supposing that this happens in the requisite form only in the modern period, with a form of systematic mathematisation, which simultaneously erodes access to the originary meaning of infinitisation, by encouraging manipulation of symbols in place of any reactivation of originary meaning fulfilments. This draws attention to two distinct notions of infinitisation and marks a shift in the attribution of responsibility for the emergence of the form of infinitisation distinctive of the so-called European sciences, that of mathematisation. Derrida clarifies the differences thus:

> Starting from this inaugural infinitization, mathematics cognizes new infinitizations which are so many interior revolutions. For, if the primordial infinitization opens the mathematical field to infinite fecundities for the Greeks, it no less *first* limits the apriori system of that productivity. The very content of an infinite production will be confined within an apriori system which, for the Greeks, will always be closed. The guide here is Euclidinian geometry, or rather the 'ideal Euclid' according to Husserl's expression, which is restricted to sense, not historical fact. Later at the dawn of modern times the apriori system will itself be overthrown by a new infinitization.
>
> (D: 1962, HOG pp. 127–28)

This then grounds the distinction drawn by Husserl between bounded and pure ideality.

For the Greeks, according to Husserl, geometrical idealities are still founded in morphological idealities. As Derrida remarks: 'The problems of origin posed outside that enclosure and concerning the sense of pre-exact or pre-objective spatio-temporality would find their place inside the new transcendental aesthetics which Husserl particularly contemplated in the Conclusion to *Formal and Transcendental Logic* (p. 291–93)' (D: 1962, HOG pp. 125–26).[15] Only then would the infinitisation of geometry cease to acquire meaning in relation to what has been thought, and instead be delimited in relation to what is thinkable. This moves the delimitation from that of bounded idealities, conditional on the activities of particular individuals, to that of an unbounded ideality, with meaning fulfilments guaranteed in virtue of their formulability alone. This transposes infinitisation from the status of a generalisation, from a number of observed instances, into a formalisation at the level of a categorial differentiation and determination. The implication is that the ideal of such a transcendental aesthetic is in turn held in place by an Idea in the Kantian sense, providing a limit conception for the infinite task of expanding geometry to a saturation of its possibilities. Up until that point, there is an arbitrariness about which possibilities within the field of geometry have been pursued by the various individual researchers, and which have not.

Derrida then puts the central claim concerning a tension between the invocation of such an Idea in the Kantian sense and the affirmation of the 'principle of all principles' of phenomenology:

> The Idea in the Kantian sense, the regulative pole for every infinite task, assumes diverse but analogous functions that are decisive at several points along Husserl's itinerary. Paul Ricoeur very precisely recognizes in the Idea 'the mediating role between consciousness and history' (R: HP p. 145). Now, while completely marking it with the highest and most constant teleological dignity, while completely granting a believing attention to what it conditions, Husserl never made the Idea *itself* the *theme* of a phenomenological description. He never directly defined its type of evidence with phenomenology, whose '*principle of all principles*' and archetypal form of evidence are the immediate presence of the thing itself 'in person.'
>
> (D: 1962, HOG pp. 137–38)[16]

Thus, while the immediate presence of the thing itself 'in person' does not have to take the form of an empirical intuition, granted the function for Husserl of categorial and eidetic intuitions, it is not obvious that there can be meaning fulfilments of either eidetic or categorial intuition for this proliferation of versions of the Idea in the Kantian sense. The problem of the sense of genesis and of the genesis of sense is thus doubled by this problem of the sense of the Idea in the Kantian sense, and of sense, as such an Idea. As an item of phenomenological enquiry, such an Idea must on each occasion be assigned a formal indication of an ideality of sense, which, through imaginative variation, may be worked on to permit the determination of the indicated essence. However, Derrida's readings reveal a deployment of the terms genesis and of the Idea in the Kantian sense in a number of different ways, the unity of which is required for the enquiries to hold together a single line of enquiry, but the formal indication of which cannot on any given occasion be deemed to have been fulfilled.

2.2 Delay, difference, *différance*

> One can desire, name, think in the proper sense of these words, if there is one, only to the *immeasuring* extent that one desires, names, thinks *still* or *already*, that one still lets announce itself what nevertheless cannot *present itself* as such to experience, to knowing: in short, here *a gift that cannot make itself present.* This gap between on the one hand thought, language and desire and, on the other, knowledge, philosophy, science and the order of presence, is also a gap between gift and economy.
>
> (D: 1991, GT pp. 29–30)[1]

In *Given Time 1: Counterfeit Money* (1991), Derrida broadens the context for a reception of Husserl's analyses of self evidence and the givenness of phenomena, into an analysis of the paradoxes of a present which is not present, of a givenness, of what cannot be anticipated, and what cannot be reciprocated, without annulling the status of the given, as gift. This shift from givenness to the gift might be thought to take the analysis out of the sphere of Husserl's concerns, since for Husserl's phenomenology, givenness is an operation in relation to securing the classical concerns of philosophy, with knowledge, and truth, its justification and its objects, and with metaphysics, as a delineation of the structures of knowledge in relation to those objects. However, once Husserl becomes committed to a thought of an incompletion underpinning the projections of idealisations, Derrida's analyses of the paradoxes of the gift can be seen as assisting in thinking the paradoxes of such givenness, as necessarily incomplete. The transition from analysis in terms of givenness, to an analysis in terms of the gift reveals a connection to an enquiry about the constitution of value, in terms of political economy, and an economy of psychic energy. Derrida invokes the structural anthropology of Mauss and of Durkheim to complement attention to Marx's more familiar account of political economy, as a critique of positive economics. Implicitly, there is a parallel from this to Husserl's critique of positivist psychology, and the development of the broader concerns of a transcendental psychology. Thus, in this later text, apparently dating from 1991, Derrida can be seen to be making connections from Husserl on givenness, and the critique of positivist psychology, to a set of concerns with restricted and general economies, in relation both to modern and to so called primitive social organisation.

However, these wider-scope notions of economy are already in play in the 1967 text, *Of Grammatology*, and the text, *Given Time*, turns out to have its origins alongside the writings going into composition of *The Post Card: From Socrates to Freud and Beyond* (1980). In *The Post Card*, it is invoked in a footnote to the essay, 'To Speculate – on "Freud" thus: 'The problematic of the "*Il y a*", (*Es gibt*, there is) was engaged in another seminar (*Donner-le temps*), fragments of which are to be published' (p. 430).[2] 'Giving time' clearly averts to the Husserlian problematic, while the invocations of the *il y a* and of the *es gibt* indicate the competing attempts by Levinas and by Heidegger to develop that problematic. The earlier seminars are adverted to again in the foreword to *Given Time 1: Counterfeit Money* (D: 1991, GT) from which the quotation at the start of this section is taken. The foreword remarks that the text is based on a series of seminars given in France, in 1977–78, and in the USA, in 1978–79, and, again, as lectures, in Chicago, USA, in 1991. The attempt is to think the relation between time, and its conditions, as masked by and masking a relation between the circulation of money, and its foundation in a distribution of wealth and reserves of entitlement. Both are caught up in processes of circulating meaning and concealing the conditions of possibility for that circulation, amongst others, shared language and continuing life.

For Derrida, if not for his audience, that text is also inflected by his own work preparing *The Problem of Genesis* for publication in 1990. Derrida describes the seminars and the 1991 text thus:

> It was in the course of this seminar that I gave more thematic figuration to a set of questions which for a long time had organized themselves around that of the gift. Was an explicit formalization of the question possible? What might be its limit? The problematic of the gift, such as it had signaled itself to me or imposed itself on me up to that point reached there, precisely at the limit of its formalization, a sort of intermediary stage, a moment of passage. The premises of this unpublished seminar remained implied, in one way or another, in later works that were all devoted, if one may put it that way, to the question of the gift, whether it appeared in its own name, as was often the case, or by means of the indissociable motifs of speculation, destination, or the promise, of sacrifice, the 'yes', or originary affirmation, of the event, invention, the coming or the 'come'.
>
> (D: 1991, GT, foreword pp. ix–x)

Consideration of the terms of the citation assists a reading of Derrida's early texts on Husserl, as well as showing how the engagement with Husserl continues into the development of these distinctively Derridean concerns. The focus is on an elaboration of the Husserlian concept of givenness, but there is a continuing interleaving of Husserlian themes with terms of analysis borrowed from Kant. The notions of formalisation and implicitly of thematisation are intertwined with those of critical delimitation and of a bounding of sense. These notions are repositioned within these broader concerns with economy and givenness, in terms of both Freudian psychoanalysis of the economy of drives, stabilising personal identity, and the Marxian concerns with the economic processes, stabilising certain divisions of wealth and empowerment. These moves shift Derrida's focus of interest away from the more narrowly drawn preoccupations with givenness, in phenomenology, into the broader concerns marking the writings of the 1970s and 1980s.

In the citation with which this chapter opens, there is an invocation of a necessity in thought of setting out as a limit case, that which cannot be made present in thought, an impossibility of determinacy which is nevertheless the condition for attempting to think at all. This limit is called 'the measureless measure of the impossible', and it echoes Heidegger's concerns in the post-war texts with the problem of measure and dimension in thinking which might render it adequate to respond to the intimations of being. The text continues:

> This gap is not present anywhere; it resembles an empty word or a transcendental illusion. But it also gives to this structure or to this logic a form analogous to Kant's transcendental dialectic, as relation between thinking and knowing, the noumenal and the phenomenal.
>
> (D: 1991, GT pp. 29–30)

Thus the explicit connection here is made not to Heidegger but to Kant, and the dialectics invoked are not those of either Marx or Hegel, but those of Kant's transcendental dialectic, from the First Critique, in which concepts and forms of argument are shown to be unreliable. The determination of this limit of thought, as the impossible, and of this impossible as a measureless measure rewrites the conception of transcendental limit, as invented by Immanuel Kant in the *Critique of Pure Reason*, to set out the scope of the deployment of concepts of the understanding such that, in conjunction with and giving form to sensory contents, there can be knowledge of objects of experience. This Kantian concern with limits is, by implication, conjoined with Husserl's analyses of presentation, as the condition for determinacy of meaning fulfilments of intuitions of sense, and with Heidegger's preoccupation in the later writings, with providing a framework or limit within which to think the implications of technology. For Heidegger, the absence of such conceptual delimitation of these forces denotes a crisis for philosophy and more generally for humankind. Thus Heidegger, the faithless disciple, takes up and develops Husserl's analyses of a crisis for European intellectual life of the lack of an understanding of its own foundations.

Derrida here signals a transition out of an analysis in terms of the categories of quantity and quality, concerning the denumerability of concepts of time, and their qualitative status, concerning an intensity of exposure to them, into a discussion in terms of the dynamical categories of relation and modality, leading into a rethinking of necessity. For Derrida, as I hope to show, the privileged place of *différance*, as one of the earliest of his invented terms, gives a priority to movement over diagrams, and to the results of deploying the categories of relation and modality over those from deployments of quantity and quality. In

Kant's terms, there is a shift here from the mathematical to the dynamical categories. Derrida develops Kant's questioning of conditions of possibility, as also a questioning of conditions of impossibility, moving from consideration of distinct conceptions of time and infinity, and their arrival in whatever modes of givenness, to a consideration of that which does not arrive, or indeed, with Heidegger, of that which precisely cannot arrive, because it withdraws in order to permit the arrival of what does arrive. This is the relation between the *Ent-eignis* and the *Er-eignis* of being. Thus Heidegger has a different critique of Husserl's 'principle of all principles': that what is most basic forms an unseen condition for that which does arrive. These are then the invisible conditions for the visible, which are to be thought of as interdependent. For Derrida and indeed for Heidegger, the giving of givenness cannot be made thematic, and the unthematisable and the thematised are interdependent, rendering what appears to have been thematised opaque.

This shifts attention to the modalities of those modes of givenness themselves, which delimit that which can and that which cannot arrive. These modalities are modes of temporality. In the following sections of my enquiries, I shall explore further Derrida's complication of the Kantian themes of possibility and impossibility, in relation to Heidegger and Levinas on death, and in relation to thinking limit and infinity, leading to a reassessment of the concept of experience. At this point, the focus is still on the continuing evocation of Husserl's problematics of givenness, as modes of temporality, and of Kant's notions of transcendental limit and transcendental illusion. The difference between the measurable, or denumerable, and the immeasurable, or incalculable, appears here briefly, if misleadingly, as a distinction between the category of quantity, enumerating distinct conceptions, and that of quality, the intensity of the evidential sources for those concepts and for the distinctions between them. The shift to thinking of time as a series of self-dislocating movements, rather than in terms of numbers of concepts of time and intensities of its registration, imposes a requirement to rethink necessity, as a delineation of necessary movement, rather than the installation of fixed reference points providing anchorage for ontology and metaphysics. In effect, Derrida's modes of reading intensify the movements of reversal and oscillation already marked up as distinctive of Husserl's and Heidegger's phenomenological enquiries to the point where there are to be found the syncopations and caesuras emphasised, by Lacoue-Labarthe, in relation to Hölderlin, and by Nancy, in relation to Kant.[3] A distinction emerges between a movement of ellipsis which rotates around a pair of fixed foci, and a movement of ellipsis, in which the foci also move.

The preceding sections have set out a series of distinctions between questions of sense, in relation to time, in relation to genesis, and in relation to history, all of which are concerns for Husserl. They have also rehearsed two further sets of concerns which arise in relation to the readings of Husserl offered by Fink, by Ricoeur, and by Derrida. For Fink, there is the question of a distinction between empirical and transcendental meaning, or between natural language and the discourse of transcendental experience, whereas Ricoeur's attention to the

Kantian components of Husserl's thinking opens out a distinction between an empirically delimited conception of experience, and a notion of transcendental experience. The latter arises as a result of Husserl's displacement of the Kantian notion of limit, in favour of his own notion of delimitation, as provided by the concept of the living present, which can also be seen to provide a concept of transcendental time. The limits within which the given intuition (*Anschauung*) concerning time can be converted into a determinate intuition (*Intuition*) of time are for Husserl provided by the conception of the living present, which, in turn, he supposes to be simply given in intuition (*Anschauung*), as it is. It would not be misleading to think of these two as related as empirical instance (*Anschauung*) to transcendental condition (*Intuition*). In this section, I shall draw together the various strands of the reading of Husserl offered by Derrida, to clarify the moves from considering time, genesis and history, to placing these considerations into the further context of the questions about meaning, experience and indeed in terms of an account of time as split between an empirical and a transcendental instance. One strand of the reading of Husserl is the multiplication of deployments of the notion of the Idea in the Kantian sense, as already set out. Another is the series of refigurings of central themes of Husserl's phenomenology, including the various claims put forward in the *Introduction to Husserl's 'Origin of Geometry'*, concerning intentionality as traditionality, sense as historicity, and the absolute as passage. The introduction of the notion of *différance* both draws these engagements with Husserl together, and provides Derrida with a route out of Husserl's mode of analysing time and meaning. Reconstructing the Husserlian backdrop out of which this notion emerges reveals its full force as a thinking of time. In the famous essay from 1968, '*Différance*', Derrida remarks a relation from its introduction to a rethinking of the relation between space and time. I shall attempt to indicate how its introduction also serves as a transition out of the reading of Husserl, on time, and indeed on givenness, while preserving some of the results of Husserl's enquiries.[4]

The remark in *Given Time 1: Counterfeit Money* indicates Derrida's diagnosis of a tendency in Husserl's phenomenology to succumb to something akin to a Kantian transcendental illusion, in a number of distinct guises. This connects back to the notion of transcendental illusion, in the preface to *Of Grammatology*, already remarked in Part I of these enquiries, entitled 'In the beginning'. This continuing use of the term 'transcendental illusion' marks the continuing working of Kant's thought in Derrida's writings. The key claim here is that, while the elements out of which an economy is composed are measurable, the gift is not reducible to a calculable order, and is the an-economic condition for any economy. The failure to observe the differences between these two resembles a transcendental amphibology, as analysed by Kant in the *Critique of Pure Reason*, in the appendix to the Transcendental Analytic, concerning the confusion of an empirical with a transcendental employment of reason. In this case, the transcendental condition, the immeasurable gift, is taken, falsely, to be empirically presented within the confines of a measurable economy. Derrida suggests that the gap opening out between the two can be taken, again falsely, to be amenable to determinate conceptualisation, and this resembles the form of the transcendental illusions analysed by Kant, in the Transcendental Dialectic, of the *Critique of Pure Reason.* This gives an analysis of how concepts of the soul, of the world and of God are taken for determinate rather than regulative concepts. By contrast to the former, the latter cannot provide determinate contents for thought, but merely provide guidelines for thinking. A further confusion between amphibology, as the confusion of status between determinate concepts of different orders, and illusion, where there is no determinate concept, is also in play. The question to pose is whether Husserl's uses of the notions of world and horizon are instances of such illegitimate conceptuality, and whether they may all the same be rescued for use in some alternate deployment.

The connection to Husserl is made by Derrida himself in the course of the text, when he introduces four lines of questioning, in terms of which he proposes to attempt to formalise the question of the gift, as a questioning of the givenness of time:

> Such an analysis can go back before speech acts, in the phenomenological style of an intentional analysis, toward the intentional act of giving, in general. On what conditions does it take place? What is a 'donating consciousness'? and so on. The latter expression, moreover, is immediately and massively complicated by reason of a figure of donation that is constantly used by phenomenologists, beginning with Husserl, to designate the ultimate recourse, phenomenology's principle of principles, namely the originary donating intuition (*gebende Anschauung*), the one that delivers up the thing or the sense themselves, in person, or in flesh and blood as people still say, in their immediate presence.
>
> (D: 1991, GT pp. 50–51)

This reference to Husserl's specification of the 'principle of all principles' of phenomenology, in *Ideas One* (section 24), is then connected up in a footnote to the transformation of that principle in the writings of Jean-Luc Marion, on God as the source of givenness.[5] This specification of a first line of questioning, with a deepening of ordinary language enquiry, in the direction of Husserl's intentional analysis, mimes the relation, for Husserl, of moving from analysis of given intentions, to that of transcendental conditions of intentionality, which may or may not be given, as present in empirical awareness.

This is immediately followed by a description of a second line of questioning, rehearsing the language of transcendental critique, and the problems of disclosing or imposing unity on phenomena, such that they may be spoken of at all, identified and re-identified, as what they are:

> One may wonder whether this multiplicity of meanings that transmits the multiplicity of *givens* and refracts it in the multiplicity of the 'to give' has a sort of general equivalent which would permit translation, metaphorization, metonymization, exchange within an ultimately homogeneous semantic

> circle. This general equivalent would be a transcendental signified or signifier. Playing the role of a transcendental given, it would orient the multiplicity and furnish the transcendental category of which all the other categories of given (to be/to have; thing/person; sensible, natural/symbolic; and so forth) would be particular determinations, metaphorico-metonymic substitutes.
>
> (D: 1991, GT pp. 51–53)

Such a general equivalence for givenness would permit an understanding of the arrival of meaning, as setting out a system of connections between spheres of thinking and spheres of objective givenness. Derrida marks up the manner in which partial object theory, in psychoanalysis, opens out this general equivalent into an incomplete series of partial gifts: '*cadeaux*, faeces, penis, child, weapons of war', and he then remarks: 'It is this problematic that we are talking about directly or indirectly' (p. 53). For Derrida, the passage to the limit, evoked by the Idea in the Kantian sense, takes the form not of a determinate, but of a non-finite series of equivalent terms, as exemplified in the given sequence of non-equivalent terms. The passage to the limit as impossibility takes the form of just such an incomplete and incompletable series of non-equivalent, and radically dissimilar individual terms. The need for indirectness in discussing this idea is given in the third and fourth lines of questioning.

The third line of questioning concerns the problem of containing the effects of idiomatic inflection of these issues, imposed by the chance formations of the natural languages in which they are discussed. Derrida then returns to question the status of transcendental enquiry, in the fourth line of questioning:

> It is thus for example that 'to give time' is not to give a given present but the condition of presence of any present in general; '*donner le jour*' (literally to give the day, but used in the sense of the English expression, 'to give birth') gives nothing (not even the life that it is supposed to give 'metaphorically', let us say for convenience) but the condition of any given in general. To give time, the day, or life is to give nothing, nothing determinate, even if it is to give the giving of any possible giving, even if it gives the condition of giving.
>
> (D: 1991, GT p. 54)

He then draws attention to a difference between this relation between condition and conditioned, and that of transcendental enquiry, by asking: 'What distinguishes in principle this division from the transcendental division it resembles?' To which he responds:

> One perceives there is no longer the sharp line that separates the transcendental from the conditioned, the conditioning from the conditioned, but rather the fold of undecidability that allows all the values to be inverted: The gift of life amounts to the gift of death, the gift of day to the gift of night, and so on.
>
> (D: 1991, GT p. 54)

and the chapter concludes by rehearsing the thought that the ambivalent status of what is given destabilises the distinction between giving and gift, while potentially providing a formalisation of this instability and of the distinctions between 'the natural and the artificial, the authentic an the inauthentic, the originary and the derived or borrowed' (D: 1991, GT p. 70). In this way, Derrida introduces the notion of undecidability as marking an inseparability of the two instances of transcendental enquiry, the transcendental condition and what it conditions or makes possible. This may not jeopardise the status of Husserl's deployment of a form of transcendental enquiry, but it is disruptive of Kant's attempt to secure the status of his version of it.

The distinction to be marked here is that transcendental illusion takes place when a concept is falsely taken to be determinate; while transcendental amphibology takes place when the wrong status is attributed to a concept. There are then three instances, in which there is some blending of the two: the first, substituting a rigorous transcendentally grounded notion of consciousness, as continuing determinacy, for intermittent processes of empirical awareness, broken by sleep, distraction, forgetfulness; the second, imposing a stabilisation of objective meaning as given for transcendental definitions, in relation to logical grammar, on the partiality and instability of natural language; and third, supposing there to be a transparency of temporal moments and a continuous series of them, as opposed to a disjunction and lack of synchrony between two distinct series: a series of punctual instants of time, construed mathematically for use in the natural sciences, and a series of empirically constituted moments, in empirical consciousness. Derrida's analyses suggest, however, that these errors of reasoning are in fact indispensable for the customary practices of human reasoning and language use. There is thus, fourth, a further potential for transcendental illusion concerning the status of history and historicality: for there is a slide between the notion of empirical, actual history, as a forwards directed series of events, and the notion of genesis, as a pre-delineated process of realising a formally given possibility, in actual thinking. The one has a future horizon which is open, while for the other, the horizon is closed. There is then a further determination of historicality, as deployed by Husserl, as the temporal conditions, in which a thought arrives for the first time, determining a distinctive set of historical possibilities. For this third notion of historicality, the horizon is radically reconstituted as a consequence of the arrival of the new configuration.

The main movement traced out within *The Problem of Genesis in Husserl's Philosophy*, and to be traced out between that text and the 1962 *Introduction to Husserl's 'Origin of Geometry'*, is that from the problem of genesis to the problem of historicality, even though in the latter text, Husserl's conception of historicality is presented more as a solution than as a problem. The distinct forms of the two texts assist in the transition, and as a result it is possible to distinguish between two kinds of transcendental enquiry: the one in which Husserl emphasises a notion of transcendental history, which is inseparably bound up with the move to genetic phenomenology, as opposed to one which moves into the mode of Kantian critical reflection, and separates itself off from questions of the fulfilments of meaning intentions. Both are detectable in Husserl's writings. Thus there is a splitting within readings of Husserl, to which Derrida's readings implicitly draw attention, between those which take this less Husserlian route, attributing to him the Kantian form of transcendental enquiry, and those which take the route of affirming a grounding of transcendental history in Husserl's analyses of genesis. Those Husserlians who refuse altogether to engage with either transcendental history or genetic phenomenology cannot then be brought into the discussion here.

In his 1953/54 preface to *The Problem of Genesis*, Derrida identifies the structure of Husserl's enquiries as following a double movement of genesis itself, in the following way:

> Now it is indeed the theme of genesis that drives all Husserl's concern, which, looked at superficially in its main methods of approach, seems to follow two vast movements, one forward, one back: first, the refusal of psychologisms, of historicisms, of sociologisms; the logical and philosophical ambitions of the natural or 'worldly' sciences are illegitimate and contradictory. In a word, the existence of a 'worldly' genesis, if it is not denied as such by Husserl, nevertheless does not attain in his eyes either the objectivity of logical meanings, or correlatively, the being or the dignity of phenomenological or transcendental consciousness.
>
> (D: 1954, PG pp. xviii–xix)

For Husserl worldly genesis, like empirical intuition, hypothesises the resulting objects as separate from that genesis, or intuition, whereas transcendental genesis, and eidetic intuition, constitute the hypothesised objects. Derrida continues:

> It is this latter which is the constituting source of all genesis; in it, originary becoming makes itself and appears to itself. The 'transcendental' reduction, end and principle of this movement, is the reduction, the farewell to every historical genesis, in the classical and 'worldly' sense of the term. But after this retreat to a philosophical purity of an idealist style, there are announced a kind of return, the outlines of a movement of broad reconquest: it is the notion of transcendental genesis, which, resistant in principle to every reduction, revealed perhaps by every reduction properly understood, will oversee a kind of philosophical recuperation of history and allow a reconciliation of phenomenology and 'worldly' sciences.
>
> (D: 1954, PG pp. xviii–xix)

The double movement is the step back from empirical genesis, through bracketing and the performances of reduction, to reveal the moves through which, for Husserl, what is encountered in empirical consciousness is constituted in transcendental consciousness, exactly through the processes of transcendental genesis.

Derrida poses three questions to this structure, and in order to pose them, it is necessary for him to impose the utmost condensation on Husserl's proliferating analyses. This, oddly, is achieved by the stylistic means of developing the three very distinct ways of reading the texts, in these three early texts: interpretation, commentary, and the alternation of decelerating/accelerating reading of *Speech and Phenomenon*. The questions are, first: is there only one sense of genesis here? Or is there here a critical vacillation, or indeed an incidence of a transcendental illusion, erasing a difference under the application of the one term, to two or more quite distinct structures, in quite distinct empirical and transcendental temporal series? The second question is: what is the nature and status of the self-presence of transcendental genesis to itself, the self-appearing of 'originary becoming': is this a punctual self-presence, within, or beyond any thinking of a time series, or does such a self-presence again conceal temporal differentiation, and a lapse of time? The third question is: is the temporality of this transcendental genesis distinct from an empirically traceable temporality of delay, deferral, non-simultaneity and of a self-divergence at an origin, in short a temporalising in a modality of movement, for which he invents the term, neither a name nor a concept, of *différance*? My preference will be to suppose that *différance* is neither one of these, neither the empirical delay nor the transcendental genesis, but rather the movement of slippage between the two. The first question mobilises a version of Kant's analysis of transcendental illusion and of transcendental amphibology to explore the processes of substituting a transcendental for an empirical question, and an empirical concept for a transcendental determination. This last distinction becomes less secure as Derrida's readings proceed, and takes on a particular role in a much later reading of Jean-François Lyotard, from 1998.[6] The second question can be seen to turn on Freudian concerns with the unpresentable conditions of possibility in childhood trauma for the currently lived identities of adult neurosis, and there is again a link from this into the later reading of Lyotard. The third question opens out the possibility that Derrida's writings all the same lie within the arc of phenomenological enquiry, not as coterminous with what Husserl writes, but understood as a series of conceptual possibilities opened out as it were in the ideal name of Husserl. The task of the rest of this section of my enquiries is to set out the trajectory of Derrida's own thinking as it emerges out of his reading of Husserl, across the rather different approaches taken to Husserl's writings in these three principal texts.

The four stages of the structural reading of Husserl in *The Problem of Genesis* mobilise a modified form of the four ambiguous conceptions of limit discussed by Kant in the Antinomies of Reason, which arise when reason is applied to the objective synthesis of appearances (A 407). These are the questions of a limited or an unlimited time series; a finite or a non-finite divisibility of matter; of determinacy or freedom with respect to events in the world; and the necessity of an absolute being. The latter two lead to a splitting of the conception of world into a world as determined by causal sequences in naturalised time and a world constituted in transcendental enquiry and as formed through the articulation of moral imperatives. An indeterminacy of argument with respect to these alternates arises, according to Kant, as a result of attempting to treat an unconditioned unity as though it had determinate conditions, which would permit determinate conceptualisation. He writes:

> We have here presented to us a new phenomenon of human reason – an entirely natural antithetic, in which there is no need of making subtle enquiries or of laying snares for the unwary, but into which reason of itself quite unavoidably falls. It certainly guards reason from the slumber of fictitious conviction such as is generated by a purely one sided illusion, but at the same time subjects it to the temptation either of abandoning itself to a sceptical despair, or of assuming an obstinate attitude, dogmatically committing itself to certain assertions and refusing to grant a fair hearing to the arguments of the counter-position. Either attitude is the death of sound philosophy, although the former might perhaps be called the euthanasia of pure reason.
> (Kant: A 407, B 434)[7]

The four Antinomies of Reason follow Kant's division of categories into the four groups: quantity, quality, relation and modality.

The first two notions of infinity, introduced by Derrida, in relation to the reading of Husserl, are those of a becoming of logic, and of an infinite totality of temporal experiences. These two pick up on the questions of the nature of a series, as open ended, but determinate, and on the nature of totality, which is unified, even if all its members cannot be assembled at any one time. They thus deploy the Kantian notions of quantity and quality with respect to analyses of infinity. The third notion of infinity is that of 'the idea of world as infinite ground' coordinating the evidences of primordial impressions as belonging to a single structure. The doubling of the given world by its infinite transcendental ground mimics the mirroring in Kant's moral philosophy of the world of determinate causalities, by a domain of non-finite human freedom. Husserl's conception of world thus by implication undercuts the strong disjunction produced between a human world of natural science, and a human and divine domain of moral implication. For Husserl, moral and factual meanings are to be analysed in terms of one and the same horizon of meaning fulfilments. Derrida identifies these three scopes of infinity, in a footnote, in their deployment in relation to three of Husserl's texts, which are understood by him thus to deploy them as Ideas in the Kantian sense:

> It was in *Logical Investigations*, vol. 1, the idea of the infinite becoming of logic; in *Ideas 1*, the idea of an infinite totality of temporal experiences; in *Experience and Judgment*, the idea of a world as an infinite ground of possible experiences. We will see how difficult it is to give a phenomenological status to these ideas that by definition precede and envelop any experience and any genesis.
> (D: 1954, PG pp. 187–88)

Derrida discusses the first in his first section, the second in his second section, and the third in the section on transcendental and 'worldly' genesis. In his

Introduction he sets out Husserl's trajectory in four stages, and introduces the last stage as centred on the question of transcendental history, as developed in Husserl's late text, *The Crisis of the European Sciences and Transcendental Philosophy*. This fourth deployment of an Idea in the Kantian sense, in the form of a concept of transcendental history, opens out the differences between Husserl and Kant on the thinking of *apriorism* and necessity.

Derrida writes of the attempt made at this stage, as transforming the concept of intentionality, and of turning the Idea in the Kantian sense into a Husserlian notion of teleology:

> To be able to reintegrate the passive genesis into an eidetic and transcendental phenomenology, the reduction and the conception of intentionality had once more to be enlarged; they had to be extended beyond the purely egological lived experience right up to an inter-subjective experience and right up to history. Once more it is an infinite idea that, in the new and more precise shape of a 'teleology' will give back an intentional sense – the only foundation of any eidetics, – to passive genesis.
>
> (D: 1954, PG p. 4)

This teleology is the arc of possible meaning fulfilments traced out by the first intimation of the thought content, and these particular local teleologies are themselves held in place by the surmise of a teleology of reason, working through the concepts of idealisation and infinitisation. In the footnote, Derrida clarifies this notion of a reiterating appeal to and introduction of a conception of the infinite: 'It is the fourth form of this idea, in the Kantian sense, that saves phenomenology from an empiricism or an existentialism (in the broad sense of that word)' (D: 1954, PG p. 187). Thus the teleological projection into the future of fulfilling intuitions of the meaning of abstractions, such as the terms of geometry, prevents the meanings of those terms from being simply identified either with what has already, as a matter of fact, been thought, which would be a form of empiricism, or with the currently constituted features of human capacity, which limits conceptuality to given human intellectual powers. This teleological notion of what is thinkable takes the place of the measure for reality provided by the notion of necessary being, as disputed by Kant's fourth antinomy. It is connected to Husserl's new notion of infinitisation, and is thus to be distinguished from the notion of teleology rehearsed by Kant.

The explicit invocation of the Idea in the Kantian sense and the implicit deployment of Kant's distinction between notions of infinity in terms of quantity, quality, relation and modality give Husserl's phenomenology and Derrida's reading of his texts the appearance of a greater congruity with the enquiries of Kant than is in fact justified. For the notions of transcendental history and of transcendental genesis take Husserl into a different domain of enquiry altogether, where transcendental conditions and empirical actuality can no longer be kept rigorously separate. The puzzle is that even though the full determination of ideal meanings exceeds any actual meaning fulfilments, nevertheless the outline of the thought must, according to Husserl, in principle have become available at one particular time, to one particular thinker. As with his invocation of Euclid and of Plato, the point is not so much that Euclid or Plato had on any one day any particular sequence of thoughts, but rather that these names are used to designate the first occurrence of human access to these ideas, whenever that first access may have taken place, whether in the life of the empirically named individual, or not. In the same way, the name 'Joyce' becomes for Derrida the name for the occurrence of 'the greatest possible synchrony with the greatest potential for buried, accumulated, and interwoven intentions within each linguistic atom, each vocable, each word, each simple proposition, in all worldly culture and their most ingenious forms' (D: 1962, HOG p. 102), even when it is possible that there are writings which exceed those of Joyce in performing such a feat.

The notion of a supplement of the origin, in the inscriptions presupposed by the activity of the voice, and the deployments of the term *différance*, provide the basis for the claim that the thinking of infinitisation made possible by *différance* is determinately finite, that is, definitively available only to empirically restricted consciousness. This thinking is made possible by interrupting the gesture of renewal, made by Husserl in response to the *aporias* recurrently encountered by him in his thinking, which rather hypothesises a non-finite series of such new beginnings. This interruption is staged by Derrida by the arrival of the name of the other, in this case the names of Kant, of Freud, of Edgar Allan Poe, and, in the *Introduction*, of James Joyce. In *Speech and Phenomenon*, the mode of reading of *The Problem of Genesis*, identifying a recurring aporetics, is deployed to interrupt the self-confirming process of an affirmative hermeneutics, which approximates to the form of reading in the analysis of 'On the Origin of Geometry'. The aporetics in *The Problem of Genesis* concerns the continual re-subordination of a thought of genesis to a thought of *eidos*. In *Speech and Phenomenon*, it concerns an aporetics in Husserl's thinking of time, which must presume both a simultaneity of temporal moments and a primacy to a duration in which meanings are sustained across lapses of attention. The priority of a living present as a self-presence of time, as simultaneous with itself, depends on the looping of time, with a retrievability of past impressions, after they have faded from recollection, and an anticipation of possible future fulfilments, even though they may never arrive, or indeed are strictly unpresentable and therefore may not arrive.

However, there is in *The Problem of Genesis* and in the *Introduction* a certain ambiguity about the relation supposed to hold between the writings of Husserl and of Kant, and there is in *Speech and Phenomenon* a certain calculated indecision between a mode of reading, in the structural form of *The Problem of Genesis*, and the mode of commentary, performed in the *Introduction*. This permits the emergence of the notion of *différance*, but it also leads to a certain obscuring of its disruptive potential. The three notions distinctive of Derrida's thought here then are those of *différance*, the various versions of undecidability, and the notion of the *survivre* as a rewriting of the notion of transcendental life. *Différance* introduces a non-Hegelian, and indeed non-dialectisable notion of difference, which makes temporal and spatial determinacy and indeterminacy part of the content of thought contents, and not merely formal conditions for thought. The thinking of the *survivre* exposes an erosion of any determinate distinction between a concept of life, in a thematisation of the living present, and a concept of death, as invoked under Freud's notion of the death drive, and as decisive for Heidegger's break with Husserl. This then undermines attempts to delimit enquiry with respect to the actual lifespans of living human beings. The notion of undecidability reveals that the two terms, in this case life and death, and time as present moment, versus time as duration, set out in a supposedly polar opposition, turn out to merge and fade the one into the other.

Derrida pushes the point yet further by suggesting that the thought of undecidability permits a merging of the enquiries of philosophy with the insights of fictional narrative. Fictional narrative in the strong sense of literature, that which studies the mark of the letter, or line, is thus a privileged site for an exploration of time as the marking of the letter, or line. The writings of Joyce are associated with an indefinable but definitive turn in the possibilities of writing, and the writings of Husserl, which are distinctively Husserlian, are then those which innovate with respect to a certain set of theoretical possibilities posed within the discipline called philosophy. The third of the epigraphs to *Speech and Phenomenon*, however, intimates a certain indistinction between the effects of the self-present voice of Husserl's originary soliloquising of meaning and the voice of Edgar Allan Poe's story, 'The Facts in the Case of M. Valdemar':

> Yes; no; I have been sleeping – and now – now – I *am dead.*

Fiction, like the gesture of bracketing, suspends naturalising assumptions, in this case about the conditions for meaningful assertion. The voice which speaks in the thematisations of transcendental phenomenology is detached from the living voice of any empirically positioned interlocutor, and has the status of a fiction, to be legitimated only by what it initiates.

The four distinct deployments of the notion of infinity, in terms of becoming, in terms of the quality of a temporal unity, in terms of a worldly horizon, and in terms of an open futurity, set out a question to the unifiability of these conceptions of infinity, which emerges even more strongly in the exchange between Heidegger and Levinas, to be discussed in Part III of these enquiries. Here it remains to indicate the connections between the readings of Husserl, the problems discussed in relation to those readings and the invention of the term, neither a word nor a concept, *différance*. It is possible now to consider the manner in which if *différance* were only a word, it could be subordinated within a Saussurean account of meaning, and if it were a concept, it could be subsumable within a Hegelian account of conceptuality. It is Husserl's phenomenology of meaning as sense which permits terms to cross over from a status as parts of a natural language, as a word, with a meaning, into a conceptual determinacy, with a sense, which may or may not be fulfilled in actual language use and thought processes. Thus the term *différance* functions within the parameters of

Husserl's enquiries, while also disrupting them and departing from them. It is thus not that Derrida performs a 'step beyond' these parameters, but that they are unstable, thus permitting *différance* to exceed them.

The expansion of the notion of intentionality by Husserl in the later writings to include inter-subjective and trans-generational transmission licenses Derrida's hypothesis in his *Introduction to Husserl's 'Origin of Geometry'*, that intentionality is transposed into a tradition and transmission of textuality. This goes in two stages, invoking first Husserl's argument concerning the sedimentation of meaning in tradition, that faded intended meanings, no longer preserved in any active retention, are preserved in written form in texts; and then the argument that trans-generational transmission takes place. The crucial claim is made in the concluding section of Derrida's reading:

> All this rigorously develops the discovery of intentionality. The latter is also nothing but the Absolute of a living Movement without which neither its end nor its origin would have any chance of appearing. Intentionality is traditionality. At is greatest depth – i.e. in the pure movement of phenomenological temporalization as the going out from self to self of the absolute of the Living Present – intentionality is the root of historicity. If that is so, we do not have to ask ourselves *what* is the sense of historicity. In all the significations of this term, historicity is *sense*.
>
> (D: 1962, HOG p. 150)

Historicality delimits the conditions in which meaning intentions may be fulfilled, and as sense is the delimitation of what can be meaningfully fulfilled, historicality, or as the translation has it, historicity is sense. This development is then thought by Derrida, in *Speech and Phenomenon*, as already implicit in the precariousness of the distinctions drawn at the beginning of *Logical Investigations*, between sign and expression, and between signitive and authentic fulfilments of meaning, through which Husserl seeks to show that there is such a thing as a fulfilment of meaning intention, in an intuition of meaning, given as the interior monologue of imaginative variation. Thus in the later text, *Speech and Phenomenon*, Derrida attributes the thought he has identified in 1962 as emergent only in the later Husserl, as already present in the earlier Husserl. This displays a commitment to the thought that there is one determinate set of philosophical commitments to be attached to the name 'Husserl', not a shifting set of surmises. Derrida supposes that intuitive fulfilment is always dependent on a supplementarity of signitive fulfilment, and that the self-present meaning of an inner soliloquy of the voice is dependent on an exterior inscription. Husserl's defence against this is simply to suppose that while in any actual case of meaning fulfilment there is reliance on information gleaned from interpreting marks on paper, that does not necessarily infect the ideal of meaning with the instabilities of natural languages and their modes of transcription.

The problem of making sense of such idealisation, without recourse to consultation of written texts, not least Husserl's own, is not the focus for Derrida's critique of Husserl's phenomenology in 1953/54, where he supposes the problem to be rather that genesis is always subordinated to *eidos*. Derrida puts the claim like this:

> The theme of transcendental genesis, which from 1919 on takes a central place in Husserl's meditation, ought to lead us back to a moment that is before any eidetics and ought to bring us close to the sphere of ante-predicative existence, of the 'life world' (*Lebenswelt*) of primitive time, of transcendental inter-subjectivity, all factors that as such are not originarily freighted with a sense arising from the activity of the 'ego'. That it seems at least is Husserl's argument. In fact we will never leave a world of constituted essence.
>
> (D: 1954, PG p. 3)

This conclusion from 1953/54 is then thrown into doubt by Derrida's subsequent writings. Indeed, a contrast between transcendental genesis and worldly genesis is thrown into doubt by Husserl himself, in the analysis of the workings of a passive synthesis, in which genesis leaves its trace, and which is not obviously worldly, in the sense of taking place in a pre-constituted world, nor obviously transcendental, in the sense of resulting from bracketing off the world. In a sense, this passive synthesis takes place before the world is formed at all, and thus in advance of its bracketability, and in advance of any distinction between worldly and transcendental analysis. Thus the distinction between worldly and transcendental analysis turns out to be a distinction for thinking, but not a distinction for what is thought, and the transposition of it, from a feature of the Husserlian methodology, through which clarification of essences may take place, into a constitutive feature of the world, in which experiences of meaning intending take place, turns out to trade on a form of amphibology. For Husserl, however, the transcendental merely parallels and reveals the constitutive features of a world, which only in a transcendental illusion appears separate from these processes of its transcendental constitution.

The infinitisation of a passage to the limit is made sense of by hypothesising what it would be to complete all the steps in the calculation, needed to provide a summation of all the terms in the series, even though in practice such a completion is impossible. This is the model for a formalisation through which to think the idealities constituting the meanings of the essences to be revealed by transcendental phenomenological analysis. In 1953/54, Derrida objects to Husserl that the notion of genesis trades on an amphibology in concepts of history; in 1962, Derrida objects that a constitution of meaning in the egological sphere in fact for Husserl depends on written transcriptions of sedimented symbols, such that reactivations of the originary insight, already distorted in its first formulation, may be achieved. In 1967, he surmises a state of neither living nor dying as distinctive of transcendental meaning, as opposed to the actual deaths of empirical subjects, and the actual omni-temporality of a concept of transcendental subjectivity. This structure of an equivalence of transcendental life as transcendental death, identified by Derrida as an intensification of living, the *survivre,* displaces the enquiry from the domain of seeking to determine numbers of steps in an argument, or numbers of items in a series, from asking how many distinct concepts of genesis, of language, of time and of memory there may be, to asking instead about the status of the auto-affection which provides Husserl with the pre-predicative evidence for this conception of transcendental life and of transcendental time, as a living present.

The notion of *différance* permits a thinking of this genesis, which is neither worldly nor transcendental. It picks up on the preoccupation with negotiating the parallel between transcendental constitution and empirical perception, on the one hand, permitting a tracking back and forth between the temporalities distinctive of empirical experience and those of the reduced time of transcendental experience. On the other hand, it also picks up on the preoccupations brought to the fore in the *Introduction to Husserl's 'Origin of Geometry'* with these various aspects of infinitisation and the various deployments of the Idea in the Kantian sense, as marked up in 1953/54. It is not misleading to suggest that indeed the notion of *différance* tracks the ambiguous event of meaning, which viewed transcendentally takes place as constitution and viewed empirically takes place as perception of externally given entities. *Différance* is then not simply the articulation together of spatial relations as also temporally articulated, but also the articulation together of distinct temporalities. The ambiguity of the term as introduced in the essay of 1968, via a consideration of a double Latin etymology, is thereby only partially disambiguated. Derrida then continues, by contrasting a Greek determination to a Latin-based etymology:

> For the distribution of sense in the Greek *diapherein* does not carry one of the two themes of the Latin *differre,* namely the action of postponing until later, of taking into account, the taking account of time, and forces in an operation that implies an economic reckoning, a detour, a respite, a delay, a reserve, a representation – all the concepts that I will sum up here in a word I have never used but which could be added to this series: *temporalizing.* 'To differ' in this sense is to temporalize, to resort, consciously or unconsciously, to the temporal and temporalizing mediation of a detour that suspends the accomplishment of fulfilment of 'desire' or 'will', or carries desire or will out in a way that annuls or tempers their effect. We shall see, later, in what respects this temporalizing is also a temporalization and spacing, is space's becoming-temporal and time's becoming-spatial, is 'primordial constitution' of space and time, as metaphysics or transcendental phenomenology would call it in the language that is here criticized and displaced.
>
> (D: 1968, SP p. 136)

This, then, has been over-swiftly understood as inscribing time within space, and space within time, instead of grasping the greater challenge of thinking incompatible notions of genesis, of historicality, and of temporalities, the empirical and the transcendental, as co-incidental, but not simultaneous, which thus requires a rethinking of space and spatiality as well.

In *Speech and Phenomenon,* the notion of *différance* is introduced two pages after the first introduction of Freud's name. This is in the fifth chapter, 'Signs and the blink of an eye', in which Derrida rehearses his reservations with respect to the claim that Husserl in *The Phenomenology of Internal Time Consciousness* has broken with the determination of time as first of all a punctual 'now' point. The text simultaneously accepts the importance for Husserl of the differences between presentation, as supported by continuous retention, and presentification, while claming that they are all the same founded in a third notion of temporal sequence, not that of the living present, but that of both continuity and discontinuity; as repetition. Derrida writes:

> Without reducing the abyss which may indeed separate retention from representation, without hiding the fact that the problem of their relationship is none other than that of the history of 'life' and of life's becoming conscious, we should be able to say *a priori* that their common root – the possibility of repetition in its most general form, that is, the constitution of a trace in the most universal sense – is a possibility which not only must inhabit the pure actuality of the now but must constitute it through the very movement of *différance* it introduces. Such a trace is – if we can employ this language without immediately contradicting it or crossing it out as we proceed – more 'primordial' than what is phenomenologically primordial.
>
> (D: 1967, SP p. 67)

This 'more primordial' challenges the supposed primordiality of originary impressions, which founds Husserl's notion of the principle of all principles. The movement of a repetition of sameness is disrupted by the invocation of a movement of iteration, which draws attention to the difference which makes it possible to identify an item as the same again but at a different time and with different conditions, and therefore as different. The contestation of priority, between Husserl's conception of the primordial, and the challenges mounted to it by Heidegger and by Levinas, and then with Derrida's responses to each, are the topic for the next part of these enquiries.

Derrida claims that the presence of presence must always be demonstrated by resort to non-presence:

> For the ideality of the form (*Form*) of presence itself implies that it be infinitely repeatable, that is re-turn as a return of the same, is necessary *ad infinitum* and is inscribed in presence itself, It implies that the re-turn is the return of a present which will be retained in a finite movement or retention and that primordial truth, in the phenomenological sense of the term, is only to be found rooted in the finitude of this retention. It is furthermore implied that the relation with infinity can be instated only in the opening of the form of presence upon ideality, as the possibility of a re-turn *ad infinitum* (*retour à l'infini*).
>
> (D: 1967, SP p. 67)

This is a turning point at infinity as much as a repetition reaching out to infinity. This double movement challenges the status of the living present as the delimitation within which empirical meaning can instantiate formally indicated sense. It leads up to the claim that the life of the living present is displaced and disputed by the notion of *différance*, imposing on it a postponement and destabilisation in the *survivre*.

> Does not this 'dialectic' – in every sense of the term and before any speculative subsumption of this concept – open up living to *différance*, and constitute, in the pure immanence of experience, the *divergence* involved in indicative communication and even in signification in general? And we mean the divergence of indicative communication *and signification in general*, for Husserl not only intends to exclude indication from 'solitary mental life': he will consider language in general, the element of logos, in its expressive form itself, as a secondary event, superadded to a primordial and pre-expressive stratum of sense. Expressive language itself would be something supervenient upon the absolute silence of self-relationship.
>
> (D: 1967, SP p. 69)

The living present will be grasped only as a posthumous reconstruction, from the other side of a reflection on death, and this in two senses: that the meaning and significance of Husserl's enquiries only begin to open up fully as a result of the efforts of his various disciples to put the various texts into the public domain, and, as importantly, only as a result of a contestation between Levinas and Heidegger on how to understand death. This turns out to conceal a prior contestation concerning death between Blanchot and Hegel, between Blanchot and Heidegger, and between Hegel and Heidegger.

When, in the following chapter, Derrida introduces the term *différance* again, and more comprehensively, he connects it up to the notion of auto-affection, and distinguishes between the movement it sets up and the constitution of transcendental subject:

> This movement of *différance* is not something that happens to a transcendental subject; it produces a subject. Auto-affection is not a modality of experience that characterizes a being that would already be itself (*autos*). It produces sameness as self-relation and self-difference. It produces sameness as the non identical.
>
> (D: 1967, SP p. 82)

Différance at this point for Derrida marks an irreducibility of a heteronomy, in advance of the self-relatedness of auto-affection.

> . . . here again we find all the incidences of primordial non presence whose emergence we have already noted on several occasions. Even while repressing difference by assigning it to the exteriority of signifiers, Husserl could not fail

> to recognize its work at the origin of sense and presence. Taking auto-affection as the exercise of the voice, auto-affection supposed that pure difference comes to divine self-presence. In this pure difference is rooted the possibility of everything we think we can exclude from auto-affection: space, the outside, the world, the body, etc. As soon as it is admitted that auto-affection is the condition for self-presence, no pure transcendental reduction is possible.
>
> (D: 1967, SP p. 82)

Thus at this stage Derrida is thinking *différance* in terms of an overturning of the status of transcendental subjectivity and of any transcendental reduction. But then Derrida indicates a necessity all the same of attempting to make the Husserlian moves:

> But it was necessary to pass through the transcendental reduction in order to grasp this difference in what is closest to it – which cannot mean grasping it in its identity, it is purity or its origin, for it has none. We come closest to it in the movement of *différance.*
>
> (Ibid.)

Oddly, it is only at this point that the English translator adds a note remarking the neologism. I shall argue in what follows that this destabilisation cannot be adequately thought in terms of these notions of otherness and heteronomy, the outside and exteriority. For once *différance* is understood as the arrival of the future, as that which can and that which cannot arrive, in the *a-venir*, and in the *a-dieu*, even this description has the effect of levelling out an unevenness of time, into a smooth temporal flow. For the *a-venir* understood as the *a-dieu* opens up a diagnosis of a rending of time between religions inscribed within onto-theology and religiosity which is not thus inscribed.

The essay '*Différance*', published in 1968, was first delivered as a lecture. It is available in two versions in English, matching the two versions published in French, one in the *Bulletin de la société française de la philosophie* (1968) and one in *Margins of Philosophy* (1972), which was translated into English and published in 1982. The first has a preface added for the published version; the second begins, as did the lecture, with the phrase:

> I will speak, therefore of a letter.
>
> Of the first one, if the alphabet and most of the speculations which have ventured into it are to be believed.
>
> (D: 1968, MP p. 3)

It is in the added preface that Derrida provides his notion with a genealogical tree:

> *Différance* is neither a *word* nor a *concept.* In it, however, we shall see the juncture – rather than the summation, – of what has been most decisively

> inscribed in the thought of what is conveniently called our 'epoch': the difference of forces in Nietzsche, Saussure's principle of semiological difference, differing as the possibility of (neurone) facilitation, impression and delayed effect in Freud, difference as the irreducibility of the trace of the other in Levinas, and the ontic-ontological difference in Heidegger.
>
> (D: 1968, SP p. 130)

This line of descent erases the very significant role which a continuing engagement with the phenomenology of Husserl plays in Derrida's thinking about time and meaning. Derrida continues to engage with the legacy of Husserl at one remove by reading the texts of Levinas and of Heidegger which do not cease to respond to and contest the inheritance of Husserl's phenomenological innovations. This provides Derrida with a different model of reading. Instead of reading Freudian thematics hidden within the Husserlian text, he has two foci for attention, the texts of Levinas and those of Heidegger given equal weight. But as both are internally unstable, this gives the form of an elliptical movement with two unstable foci. I shall turn now to a reconstruction of Derrida's readings of Levinas with Heidegger, of Heidegger with Levinas, and of each as disrupted by the readings of Blanchot, in which a different kind of interruption of the coherence of one order of enquiry, by the arrival of the name of the other, is to be traced out, and in which the name 'Husserl' remains as a scarcely erased trace.

Part III

Experience and limit

Heidegger, Levinas, Blanchot

3.1 'A time that has always already run out'

> The question-prayer that turned me toward him perhaps already shared in the experience of the *a-Dieu* with which I began. The greeting of the *a-Dieu* does not signal the end. 'The *a-Dieu* is not a finality', he says, thus challenging 'the alternative between being and nothingness', which is not ultimate'. The *a-Dieu* greets the other beyond being, in what is 'signified, beyond being, by the word "glory"'. 'The *a-Dieu* is not a process of being: in the call, I am referred back to the other human being through whom this call signifies to the neighbour for whom I am to fear.'
>
> (D: 1996, AD p. 13)[1]

The citation shows Derrida wrapping his response to Levinas around fragmentary citations from Levinas' essay 'Bad Conscience and the Inexorable', first published in 1981.[2] The citation is taken from Derrida's *Adieu: To Emmanuel Levinas*, which contains his funeral address, and the long essay 'Word of Welcome', contributed to the conference in 1996, 'Homage to Levinas', in which Derrida discusses the politics of hospitality and the status, for Levinas, of the subject as taken hostage for the well-being of the other. Levinas' essay, in turn, carries at its head a citation from Blanchot's *The Writing of the Disaster* (1980).[3] In that text by Blanchot, there is to be found the following remark concerning time and the other:

> The other is related only to the other: the other repeats, but this repetition is not a repetition of the same; the other redoubles by dividing and diverging infinitely from himself, affirming a time outside of any future, present, or past (a time which the other thus negates) – a time that has always already run out.
>
> (B: WD, p. 34)

This otherness resists the move made by Husserl of supposing that thinking and meaning refer themselves back to a self-sameness, guaranteed at a transcendental level, underpinning the continuity of empirical selves and identities.[4] The nesting of a text by Blanchot within a framing provided by an essay published by Levinas, and in turn within Derrida's reflections on the death of his friend, sets up a distinct set of problems for commentary, discussion and analysis, for which an impossible temporality, that of simultaneous readings of several texts, is required.[5] The time which has already run out, indicated by Blanchot, is akin to the time of living on, ghosting the time of past, present and future, but no longer delimited in relation to a natural or historical sequence of time and tense. It is this time that the complicated relation of mutual reading, also called friendship, between Blanchot, Levinas, Derrida and their continuing points of reference: Aristotle, Augustine, Nietzsche; Hegel, Husserl, Heidegger, both intimates and renders the more paradoxical and aporetic.[6]

The time of the disaster is for Blanchot a time which might be registered only after the event, and is therefore strictly speaking not registered at all. Occurrences in a mode of temporality which cannot be registered are in effect of a kind which cannot be experienced, and which therefore cannot be supposed amenable to a direct seeing, of a Husserlian type, through which for Husserl both givenness and evidence are secured.[7] In this same text, Blanchot's *The Writing of the Disaster*, the following is also to be found:

> The disaster does not put me in question, but annuls the question, makes it disappear – as if along with the question, the 'I' too disappeared in the disaster which never appears. The fact of disappearance is precisely not a fact, not an event; it does not happen, not only because there is no 'I' to undergo the experience, but because (and this is exactly what presupposition means), since the disaster always takes place after having taken place, there cannot possibly be any experience of it.
>
> (B: WD, p. 28)

There are two distinct movements of disruption here. Heidegger's interrogation in *Being and Time* (1927), of a failure to pose the question of being, in the course of the history of philosophy, is disrupted by the parallel invocation of a response, in prayer, to a voice or call which falls outside the scope of philosophical thematisation.[8] Second, the event (*Ereignis*), as resolution, or *Entschlossenheit*, in Heidegger's analyses of time, in *Being and Time* (1927), and as recast in the later *Of the Event: Contributions to Philosophy* (1938), is paralleled by Blanchot's invocation of an annulment of questioning in a 'non-happening' of disappearance.[9] The latter disrupts the unification of time attempted by Heidegger, in terms of temporal ecstases, in *Being and Time* and, subsequently, in terms of the thinking of the event.[10] It is the disrupted temporality of thematisation and disruption of thematisation that these mutual readings of Derrida, of Levinas and Blanchot present to view. This 'worn out time' of the impossibility of synthesis is already in play in much earlier texts by Blanchot, *The Step (not) Beyond* (1973) and *The Infinite Conversation* (1969).[11] Towards the end of this section of the discussion, I shall turn to an even earlier formulation of this in his essay 'Literature and the Right to Death', from 1948.[12] These various writings are responded to by Derrida in the formation of his notion of 'living on' (*survivre*) discussed in relation to a reading in the marginal commentary on Blanchot's *récits* in 'Living On – Border Lines' (*Survivre: Journal de bord*) of 1978[13]. The discussion suggests that thematisation is always parasitic on and derivative from a discursive formation which defies thematisation, and in which determinate meaning does not arrive.

In this chapter then I trace out the refusal of a thematics of time enacted by Derrida in *Aporias: Dying – Awaiting (One Another at) the 'Limits of Truth'* (1993), and indicated in this other thinking of time, which can be registered only in the modes of passivity, as written about at length by Blanchot.[14] The citation from *Adieu* intimates that, through this encounter with Levinas, Derrida underlines this doubling of Heidegger's emphasis on the question, already discussed by Derrida in *Of Spirit: Heidegger and the Question* (1987).[15] The Levinasian prayer performs an address to the God, to whom Derrida here dedicates his memories of Levinas, as he bids his friend goodbye, or, in its unforeshortened form, God be with you. Thus Derrida explores the ambiguity of bidding someone goodbye, commending them to God, and addressing oneself in prayer to one's God. The expansion of the expression 'goodbye' into its etymological ancestor indicates a suspicion to be explored here, concerning the retention, within common sense and common language expressions, of metaphysical and ontological commitments concerning theology and religion. The notion of experience also goes through a process of doubling, with a displacement, from the scope confined within the limits of a 'metaphysics of presence', as diagnosed in 'Violence and Metaphysics: An Essay on the Thought of Emmanuel Levinas' (1963), through the intensive disruption of any such notion of experience, in the time of a passivity beyond all passivity.[16] The notion of experience shifts towards one through which a notion of limits itself might be constituted, in the movement of the passage enacted in experience. This contrasts to the Kantian notion of experience, as given in the *Critique of Pure Reason* (1781, 1787), which takes place within the limits set out by a prior analysis of a connection between space and time, as given according to Kant, in intuition.

This section opens out a complication in the notions of both limit and infinity, and indeed of experience, as staged here by Derrida's response to Levinas. For Derrida, against Levinas, seeks a notion of a passage to and from the limit, which can permit a thinking of infinity, neither assuming, nor ruling out the arrival of an infinity, thought as arriving from a divine source. Levinas disputes this possibility, as the passage from 'Bad Conscience and the Inexorable' in its complete form suggests:

> Infinity would have no meaning for a thought that goes to the limit, and the *a-Dieu* is not a finality. This is perhaps what the word 'glory' signifies beyond being: the irreducibility of the *a-Dieu*, or of the fear of God, to the eschatology by which, in the human, the consciousness is interrupted which went toward being in its ontological perseverance, or toward death which it takes for an ultimate thought. The alternative of being and nothingness is not ultimate. The *a-Dieu* is not a process in being. In the call to me, I am referred to the other human being, through whom this call signifies, to the neighbour for whom I have to fear.
>
> (L: GCM, pp. 176–77, trans. mod.)[17]

Thus Levinas intimates that 'glory' may signify a sense of divinity, beyond any eschatological figuring, and certainly beyond ontological determinations of conceptions of God. This 'call to me' is marked out in its movement as quite distinct from and contrasted to the call of the self to itself, hypothesised by Heidegger as the moment at which fallen Dasein, in the mode of disowned existence, is retrieved, and Dasein affirms itself as being a self.[18] The call from on high stands as a challenge to the analysis of the self-address of the call of conscience, in Heidegger's analytic of Dasein. The one call arrives from beyond a surmised limit, whereas the other call, hypothesised by Heidegger, plays a role in instituting a limit. Heidegger writes:

> Indeed the call is precisely something which we ourselves have neither planned nor prepared for nor voluntarily performed, nor have we ever done so. 'It' calls, against our expectations and even against our will. On the other hand, the call undoubtedly does not come from someone else who is with me in the world. The call comes from me and yet from beyond me.
> (H: SZ, p. 275; MR, p. 320)

and Heidegger concludes with a question: 'What if this Dasein which finds itself in the very depths of its uncanniness should be the caller of this call of conscience?' (H: SZ, p. 276; MR, p. 321).This call, calls the 'me' into self-nomination, whereas the Levinasian call reserves the moment of naming for God.

Both Levinas and Heidegger set out a relation to Husserl's separation of a notion of fulfilments of meaning, from that of a fulfilment of time. Neither Heidegger nor Husserl supposes that a fulfilment of meaning requires a fulfilment of time, either as divine redemption or as teleological completion, and certainly not as both. However, once an eschatology of a fulfilment of meaning is deemed to arrive out of another kind of infinity, distinct from that of the infinitisation surmised by Husserl, there is a need for a discussion of the changed status of this notion of fulfilment, and of the status of this other notion of the infinite, and their relation to any such fulfilment of time. The move from surmising this wholly other infinity, to surmising it to take the form of a determinately transmitted experience of the divine, in a specific tradition of religious practice, is a further move, the implications of which appears to limit the universal transcendental scope of Husserl's analyses, which is neutral with respect to the form of religious commitment. There is then an irreducible contestation between Emmanuel Levinas and Jean-Luc Marion about which tradition of religious practice is to be endorsed. In the revised version of *The Gift of Death* (1992, 2003) and as marked up by Gil Anidjar, Derrida disrupts this contestation by invoking a third, by way of the name Ishmael, and a reception of Islam.[19]

The suspicion concerning the concept of experience is voiced thus, on the penultimate page of Derrida's essay 'Violence and Metaphysics', again citing Levinas, this time from *Totality and Infinity: An Essay on Exteriority* (1961)[20]:

> The experience of the other (of the infinite) is irreducible, and is therefore 'the experience *par excellence*' (TI). And, concerning death which is indeed its irreducible resource, Levinas speaks of an 'empiricism which is in no way a

positivism'. But can one speak of an *experience* of the other or of difference? Has not the concept of experience always been determined by the metaphysics of presence? Is not experience always an encounter of an irreducible presence, the perception of phenomenality?

(D: 1967, WD p. 152)

The route out of this difficulty is to hypothesise a form of phenomenology without presence. This leads Derrida in *Specters of Marx* (1993) to hypothesise the phenomenon of spectrality, and to double a phenomenology linked to ontology by a phenomenology linked to a hauntology, of analysing appearances which do not presuppose presence. It is for this reason that Derrida invokes Blanchot in that text, since neither Marx nor Blanchot are constrained by the realism of Husserl's analyses of phenomenality.[21] However, there is an alternative account of Husserl's phenomenology which would oppose an analysis in terms of perception, and as privileged by Merleau-Ponty, to an analysis in terms of phantasy, and image consciousness, and would assign the privilege to the latter.[22] If the most basic moves in constitution are to be thought rather in terms of image consciousness than in terms of perception, then the Derridean analyses of spectrality are not so far removed from those of Husserl after all.

Suspicion concerning the status of an experience of infinity, of the other, or of difference motivates Derrida's affirmation of an undecidability between three or more incompatible and incommensurable thematisations of time. This affirmation is to be found in the text *Aporias*, where the incompatibilities of the analyses of time proposed by Heidegger and Levinas, by Blanchot and Freud are invoked. What is odd is the absence of return to the analyses of time offered by Husserl. The disjunctions between these various accounts of time set up a gap between a contested metaphysics of presence, proposed as philosophical task, and an unquestioned, unquestionable metaphysics of presence, transmitted implicitly. For this text there is neither one concept of experience, nor one undisputed, indisputable concept of a metaphysics of presence, nor one unified, unsyncopated temporal horizon. In the earlier essay, 'Violence and Metaphysics', a suspicion is voiced against transcendental philosophy, and against a unitary conception of a 'living present':

In the last analysis, if one wishes to determine violence as the necessity that the other not appear as what it is, that it not be respected except in, for and by the same, that it be dissimulated by the same in the very freeing of its phenomenon, then time is violence. This movement of freeing absolute alterity in the absolute same is the movement of temporalization in its most absolutely unconditioned universal form: the living present. If the living present, the absolute form of the opening of time to the other in itself, is the absolute form of egological life, and if egoity is the absolute form of experience, then the present, the presence of the present and the present of presence, are all originarily and forever violent. The living present is

> originally marked by death. Presence as violence is the meaning of finitude, the meaning of meaning as history.
>
> (D: 1967, WD p. 133)

This strongly links a certain notion of time and its temporalisation to a notion of violence, via the privileging of a conception of a living present, for which egological life is presumed necessarily to be complete in itself, taking in each instance the same form. If, however, egological life is marked internally by alterity, and even by mutually non-communicating levels, that life is marked by a syncopated temporality, in which the various levels in that egological life are not subordinated to each other. This hypothesis is figured by Derrida in *Speech and Phenomenon* through the invocation of the name 'Freud', but it might be thought to be already available in readings of Husserl which loosen the connection between individual empirical egos and the constitution of transcendental subjectivity. Presence as violence, and transcendental philosophy as committed to that violence, then become necessary results of only one interpretation of the transcendental task and of transcendental subjectivity, as outlined by Husserl.

It thus becomes important to consider whether in Husserl's thinking, the layers of empirical and transcendental subjectivity, and of active and passive consciousness can permit of a radical alterity, preventing these closures. Respect for alterity would require respect for incommensurable temporalities, presenting distinct sequencings of time, and preventing simultaneous presentations of same and other, of self and other in me, in one determinate living present. For such deconstruction, there is no passage from a critique of the various thematics of time, to doctrine. The question would then be whether critique, as inaugurated by Kant, can accommodate or must resist this disruption, if various levels of time and temporalisation may be distinguished, but not systematically linked up to one another and articulated as systematic unity. The readings given in Part II suggest that Husserl's analyses begin to open out such a possibility. For Husserl, the meaning of meaning is not history, which is rather a process of covering up meaning, in the sedimentation of transmission, but a process of constituting history. Thus history as that which happens and as given sequence is interrupted by bracketing, and by the marking of meanings in their moments of historical inauguration: their historicity. Thus historicity runs contra to a naturalised notion of history and indeed to a teleological notion of history as completable. For Husserl, a fulfilment of meaning is separated off from any thought of a fulfilment of time, either in a divine redemption, or in a teleological completion. This then sets up a gap between the theories of meaning offered by Husserl, and that offered by Hegel, for the latter does not distinguish in the same way between the two modes of fulfilment, of meanings and of time

Levinas separates off eschatological fulfilment, from a fulfilment of time as historical series by insisting on a break in time between three temporal moments: a historical and totalised duration, the dead time of its rupture, and the fulfilled time of creation, which may or may not come again in a messianic break of historical, totalised duration. This rehearses the structures of messianic fulfilment as set out by Franz Rosenzweig in *The Star of Redemption* (1921), as opposed to the fulfilment of history as analysed by Hegel.[23] This then constitutes a break with Hegel, but it may not constitute a definitive break with Husserl. Levinas is inclined to affirm the eschatological break of an arrival of the divine in human lives, but to refuse to link it to any teleological completion of a divine will in a realised state of affairs in the world. He writes in *Totality and Infinity*:

> The interval of discretion or of death is a third notion between being and nothingness.
>
> This interval is not to life what potency is to act. Its originality consists in being between two times. We propose to call this dimension dead time. The rupture of historical and totalized duration which dead time marks, is the very rupture that creation operates on being.
>
> (L: TI, p. 58)

Here again is the emphasis on an interval separating being from nothingness, stalling Hegelian mediations, and suggesting that Sartre's ontology and Sartre's Husserl is the target of this critique, rather than Husserl himself. A doubt about a connection between divine time and a future, historical time is mirrored in the title essay of Blanchot, *Le livre à venir* (1959).[24]

There, the future is sketched as the time of the arrival of the literary writing, which does not set up a relation of representation to a world beyond itself, but which rather sets out the limits within which a representation can take place. Blanchot's vision of the space of literature is one strand of a challenge to Husserl's attempt to set up a single all-encompassing account of the limits of meaning. For Blanchot's writings imply that literary activity constitutes a series of multiple singular horizons for meaning fulfilment, in the invitation to reading. Levinas' insistence on opposing totality and infinity, and on opposing the two times of secular history and of a divine eschatology, is another strand of this challenge. Each, the invocation of a space of literature, and the invocation of divinity, sets out distinct and incompatible determinations of time and of its genesis, in temporalisation. What is peculiar about Derrida's response to this is that, with some assistance from conceptual resources provided variously by Husserl, Heidegger, Kant and Jean-Luc Nancy, he inaugurates a third thinking of time as undecided between the two: literature and divinity. In *Aporias: Dying – Awaiting (One Another at) the 'Limits of Truth'*, an invocation of the name 'Blanchot' prevents an exclusive dialogue between Heidegger and Levinas from dominating the line of argument. A further invocation of the name 'Freud', through which Derrida introduces a notion of originary mourning, undercuts any irreducible distinction between self and other, and thus prevents the views of Heidegger and Levinas, on the priority of the self and of otherness, from taking on the form of an antinomy.[25]

In the first section of these enquiries, I indicated that Heidegger and Levinas respectively seek to appropriate and complete Husserl's enquiries by rethinking the horizon for the determination of meaning, as thought by Husserl in the living present. In this part of my enquiries I shall show how Derrida explores the ways in which Heidegger and Levinas, and, in addition, Blanchot, provide variants of a notion of a passage to the limit, which set out alternate accounts of this horizon, and of its constitution. The concept of a horizon is important, then, since it provides a delimitation for the scope of the determination of meanings, and it is strongly related to temporalisation, as the processes determining the modes of givenness of that to which meanings attach. For Husserl, the time of the living present permits a thinking together of the three aspects of time which he identifies for attention: natural time, as subject to naturalisation and to naturalistic fallacies; temporalisation, as the modes of givenness of distinct domains of entities, the temporality or *Zeitlichkeit* of these domains; and the genesis, or constitution of time itself, called temporising, *Zeitigung*. These are revealed as possible objects of thought through repeated performances of the sequences of reductions, whereby consciousness is led to reveal itself as, at core, an absolute flow and flux of temporalising.[26] How precisely these four aspects of an account of time, and of its thinkability, are to be linked together is what is in dispute when Levinas and Heidegger, and in addition Blanchot, put in question the horizon within which Husserl seeks to think the relation between them.

Derrida's readings of Husserl reveal how the notion of genesis plays a crucial role in permitting a thinking of a distinction between time as naturally given sequence and time as an aspect of a process of transcendental constitution. However, the notion of genesis itself defies the requirement of presenting itself in an intuition, which can be fulfilled and delimited with respect to the expectation structures of retention and protention. In *The Problem of Genesis*, Derrida supposes that the open-endedness of a process of genesis is foreclosed by the requirement that its meaning be given formally, in advance, as formal indication. This is subsequently transformed by him into the problem and paradox of invention, in the essay '*Psyche:* Inventions of the Other' (1987):

> It is certainly expected of a discourse on invention that is should fulfil its own promise or honor its contract; it will deal with invention. But is it also hoped (the letter of the contract implies this) that it will put forth something brand new – in its words or its contents, its utterance or its enunciation – on the subject of invention. To however limited an extent, in order not to disappoint its audience, it ought to invent.
>
> (D: 1989, *Psyche* p. 27)[27]

Later in the same essay Derrida continues to articulate this structure of invention, in relation to time, and to an opening up of the future:

> Passing beyond the possible, this *différance* or writing without status, without law, without a horizon of reappropriation, programmation, institutional legitimation, it passes beyond the order of demand, of the market for art or science, it asks for no patent and will never have one. In that respect it

> remains very gentle, foreign to threats and wars. But for that it is felt as something all the more dangerous.
>
> Like the future. For the time to come is its only concern: allowing the adventure or the event of the entirely other to come. Of an entirely other that can no longer be confused with the God or the Man of ontotheology or with any of the figures of the configuration (the subject, consciousness, the unconscious, the self, man or woman and so on.) To say this is the only future is not to advocate amnesia. The coming of invention cannot make itself foreign to repetition and memory. For the other is not the new.
>
> (D: 1989, *Psyche* p. 61)

The coming of invention as the *a-venir* may not be foreign to repetition and memory, but it precisely transforms them into the notions of iteration and *restance*, which are structures of retrieval and remainder not inscribed within a concept of spontaneous, self-present empirical consciousness. The structure, iteration and *restance*, articulates in one way a certain duplicity in the notion of *différance*, which Derrida seeks at one and the same time to leave undecided, and to set out in such a way as to reveal a possibility of thinking, without privileging spontaneous empirical acts of consciousness over receptivity and passivity, and without privileging intentionality over the processes of sedimentation and of the genesis of consciousness. This thinking turns out to be grounded neither in literature, nor in theological argument, but in the hybrid form of confession, which threatens the security of the distinction between the two.[28]

Husserl's conception of the horizon is understood by Derrida to foreclose the future, as programmed in advance by what has already occurred, in a protention founded in retention. This, however, aligns Husserl's concept of the future and of the living present too closely with the formalism of Kant's conception of time and space and aligns Husserl's deployments of the notion of the Idea in the Kantian sense too closely with the closed horizon of their deployments in the *Critique of Pure Reason*, rather than grasping the possibility of a different construal of such formalism already made available in Kant's *Third Critique*, and seconded in Husserl's formalism.[29] For these, formalism is by no means to be opposed to a materialism, and there is even a place for material modification, to drive the process of invention. This alternative account of Husserl's conception of the future is opened out for Derrida by Levinas' reading of Husserl, which refuses the privilege assigned to the concept of intuition, and in its place installs the conception of the face, as a rewriting of the notion of the living present, installing as well a concept of absolute experience, as a challenge to the notion of conditional experience, as set out in Kant's critical system. This Levinas performs emphatically in the opening sections of *Totality and Infinity: An Essay on Exteriority* (1961), where he writes:

> Form – incessantly betraying its own manifestation, congealing over into a plastic form, for it is adequate to the same – alienates the exteriority of the

> other. The face is a living presence; it is expression. The life of expression consists in undoing the form in which the existent, exposed as a theme, is thereby dissimulated. The face speaks. The manifestation of the face is already discourse.
>
> (L: TI, p. 66)

Thus Levinas affirms the expressive moment of discursivity over the indicative function, thus apparently affirming the priority argued for by Husserl in the first of his *Logical Investigations*, while displacing the account of meaning away from the postponed fulfilments of a formal indication, in favour of a more immediate givenness, not of intuition but of expression as a form of revelation. The immediately preceding sentences are:

> Here, contrary to all the conditions for the visibility of objects, a being is not placed in the light of another but presents itself in the manifestation that should only announce it; it is present as directing this very manifestation – present before the manifestation, which only manifests it. *The absolute experience is not disclosure but revelation*; a coinciding of the expressed with him who expresses, which is the privileged manifestation of the Other, the manifestation of a face over and beyond form.
>
> (L: TI, pp. 65–66)

Thus the exteriority for which his intervention seeks to make way is one not just of otherness as a heterogeneity within a given set of spatio-temporal dimensions, but introduces an alternate set of temporal determinations, and the conception of an interval, or caesura between the two time sequences.

The thinking of the face is thus directly linked to the primacy of a concept of exteriority, and to its derivation from an experience, which reveals a divine intending, otherwise called creation. Levinas continues his distancing from Husserl and indeed from Kant, in the following way:

> Signification is not an ideal essence of a relation open to intellectual intuition, thus still analogous to sensation presented to the eye. It is pre-eminently the presence of exteriority. Discourse is not simply a modification of intuition or of thought, but an original relation with exterior being.
>
> (L: TI, p. 66)

This meaning, Levinas continues:

> is not produced as an ideal essence; it is said and taught by presence, and teaching is not reducible to sensible or intellectual intuition, which is the thought of the same … it is a presence more direct than visible manifestation, and at the same time a remote presence – that of the other.
>
> (Ibid., p. 66)

That this is also a critique of Kant becomes more clear when it is put in the context of the move made on the previous page with respect to Kant's account of time. Levinas is considering the connections between history, pastness and knowability:

> For to know objectively is to know the historical, the fact, the already happened, the already passed by. The historical is not defined by the past; both the historical and the past are defined as themes of which one can speak. They are thematized precisely because they no longer speak. The historical is for ever absent from its very presence. This means that it disappears behind its manifestations; its apparition is always superficial and equivocal; its origin, its principle, always elsewhere. It is a phenomenon – a reality without reality. The flow of time in which, according to the Kantian schema, the world as constituted is without origin.
>
> (L: TI, p. 65)

Derrida's objection to this is not so much that it is a departure from either Kant or Husserl. Indeed, he affirms the critique of this aspect of Kant's account of time and space, and of the implied conception of experience, as inscribed within a metaphysics of presence; and he is in sympathy with the distancing from Husserl's concept of intuition.

However, the move is from a temporal postponement of an arrival of meaning, with Husserl, in favour, apparently, of an unequivocal arrival in the present moment, of an intimation of divinity in the face of the other, the very opposite of the direction in which Derrida is inclined to take Husserl, which is rather towards the claim that there is no fulfilment of meaning. This temporality of immediacy in Levinas is a focus for critique in Derrida's paper 'At this Very Moment in this Work Here I Am … ', first published in 1980, responding to the modified eschatological moves made in *Otherwise than Being: Or Beyond Essence* (1974), as 'Violence and Metaphysics' (1963) had responded to *Totality and Infinity*.[30] Derrida thus continues to be wary of Levinas' affirmation of an eschatological moment, in advance of history, as affirmed thus in the preface to *Totality and Infinity*:

> The eschatological, as the 'beyond' of history draws beings out of the jurisdiction of history and the future; it arouses them in and calls them forth to their full responsibility. Submitting history as a whole to judgment, exterior to the very wars that mark its end, it restores to each instant its full signification in that very instant: all the causes are ready to be heard. It is not the last judgment that is decisive, but the judgment of all the instants of time, when the living are judged. The eschatological notion of judgment (contrary to the judgment of history in which Hegel wrongly saw its rationalization) implies that beings have an identity before eternity, before the accomplishment of history, before the fullness of time, while there is still time; implies that beings exist in relationship, to be sure, but on the basis of themselves and not on the basis of the totality.
>
> (L: TI, preface p. 23)

This differentiation between Hegelian judgment and eschatology, and that of Levinas, opens up a gap between a conception of eschatology, breaking with any natural teleology and summation of history, and a concept of eschatology which, with Hegel, imposes a refinement and completion on natural and historical process.

While treating these eschatological themes with caution, rehearsing some of his reservations in the famous paper 'Of an Apocalyptic Tone Recently Adopted in Philosophy' (1980), Derrida in effect borrows from Levinas the conception of interruption, of the *a-dieu* and of an immemorial past, out of which the thought of divinity arrives, in order to install a conception of a future which is not foreclosed, within what he understands to be the restrictions of sameness implicit in the Husserlian and Heideggerian conceptions of the horizon and of futurity. The notion of the *a-venir* which comes increasingly to the forefront of his thinking is a modification of the *a-dieu* affirmed by Levinas, the directedness towards a divine revelation, which is prompted in advance by a resonance with the immemorial past of creation. He deploys Levinas' own notion of the interval to hold open as undecided the competing accounts of the primordiality of time as finite, or as infinite, which connects to the notion of interruption. This allows the thought of Blanchot and indeed that of Husserl to prevent the dispute between Levinas and Heidegger concerning time from exhausting the options concerning the thinking of time. The resulting differential thinking of futurity and of the trace, and the different deployments of the notions of death and finitude, open out Derrida's distinctive concept of the *a-venir*, as a displacement of the remaining religious commitments of the Levinasian '*adieu*'. There are, however, further strands of religious and theological thinking contributing to Derrida's thinking of time, which will require discussion.

For Heidegger the thinking of futurity opens up as a result of posing a need to retrieve what has gone missing in the transmission of an originary insight about meaning and being. This opening up of the future is then rethought by both Levinas and Derrida. Thus while as frequently insisting in the gaps between his own thought and that of Heidegger, Derrida traces a close connection between Heidegger's notions of destruction and appropriation, and his own invention, deconstruction. Thus he writes in the fourth paper on *Geschlecht*, 'Heidegger's Ear: *Philopolemology*', (D: 1989, G4): 'Deconstruction, or rather *Destruktion*, is also an experience of the appropriation of tradition' (p. 180). The question to pose, however, is whether this is experience, as an occurrence within a horizon of meaning, or an experience as constituting a horizon of meaning. As the latter, deconstruction comes closer in meaning to that of Husserl's analyses of the movements of genesis, sedimentation, and reduction, which reveals an original, historically given genesis. This way of responding, however, reinstalls a Husserlian distinction between empirical and transcendental experience, between empirical intuitions, as providing access to externally given meaning contents, and transcendental constitution, that out of which those empirically given contents arise in the first place. It is thus all the more important to distinguish between the transcendentalism of Kant and that of Husserl, for making appeal to that of Husserl may not necessarily imply a return to the claims of Kant's *Critique of Pure Reason*.

It is important to note that Derrida's readings of Husserl serve to detach Husserl and indeed Kant from the stabilisations of scope of meaning in the First Critique in favour of a recognition of an *apriori* marking of the scope of concepts by their historico-empirical origins. For Husserl's transcendental constitution takes place in historical time, as perhaps does that of Kant's Third Critique. Thus there is a violence in supposing there to be only one account of the transcendental. Similarly, a return to Husserl is not necessarily a return to Heidegger, and vice versa. The problem for Husserl's notion of genesis is that it is so basic to the mode of analysis that it is hard for there to be a failure of adequation for its protentional expectation, without a complete collapse of meaning. Since it is constitutive for the scheme of thinking, it is hard to conceive of intuitions concerning it being shown to be empty. Heidegger's reading of Husserl transposes the question of genesis and of the genesis of sense into the registers of the analysis of tradition in the sendings, withdrawal and forgetting of being, which can be revealed to view only by performing a destruction of the inheritance, on an analogy with Husserl's notion of reduction. For the Heidegger of *Being and Time*, there is meaning for Dasein because Dasein is positioned in time, with a relation to time; and that meaning has historical determinacy because of that positioning, or situatedness of the *da*, of Dasein, which accompanies any intimation of being.

For the Heidegger of the thinking of *Ereignis*, by contrast, there is meaning within an epoch of the sending of being because historical composition, with Hölderlin and Nietzsche, and perhaps with Aristotle and Anaximander, bends language into a decisive figuration, opening an insight, or indeed intuition into the connections between time, meaning and its genesis as history. The appropriation of being into meaning constitutes an epoch in parallel to the processes whereby Dasein appropriates meaning to itself as a self through its grasp and understanding of its inheritance and of its future, as opened out in its capacity for being. As shown by Kisiel in his discussion of the development of *Being and Time*, already in advance of the writing of *Being and Time*, Heidegger had considered the temporality of the *Ereignis*, the event as arrival of determinate, appropriable meaning.[31] This is initially thought by Heidegger in the biblical terms of the call to become an apostle or disciple, and in terms of the arrival of the gift of multi-lingualism in the event of Pentecost (*Acts of the Apostles* 2.4):

> And there appeared to them flames like tongues of fire distributed among them and coming to rest on each one. They were filled with the Holy Spirit and began to talk in other tongues as the Spirit gave the power of utterance.

This is traced out in outline in the notes gathered together in the published version of the lectures from 1920–21, *Phenomenology of Religious Life* (GA 60: 1995). In the later texts, Heidegger attempts to rethink such an originary arrival of meaning through less obviously religious sources, the poetry of Sophocles, Hölderlin, Trakl. This poses the problem relation of an overlap of scope between literature and religion, as charting the limits of articulable experience, which is to be taken up more directly in the next part of these enquiries. For Derrida, each serves to circumvent the immobilisation of such questions in the gestures of onto-theology, which suppose that these limits have been settled once and for all time in a divine creation, coextensive with the possibilities of meaningful articulation of a *logos*. Thus for Derrida each, literature and religious inheritance, are forms of resistance to the imposition of a false universality through ontology.

Husserl's notion of historicity is tied to the notion of genesis; Heidegger's notion of historicality is tied to the notion of the sendings and withdrawals of being. Thus the difference between Husserl's notion of historicity and Heidegger's notion of historicality is that for Husserl there is historicity not as a result of the sendings of being, but because there must always have been a first occasion on which any meaning content has been thought. That first occasion may not as a matter of fact be retrievable, and indeed need not be retrieved, since that first occasion is implicitly contained within any other thinking of the same content. The problem for Husserl comes with the re-identification of contents as 'same', when, for example, with the thinking of infinitisation, between the time of Euclid and the time of non-Euclidian geometries, the notions of finitude and infinity appear to have significantly shifted. For Heidegger, historicality attaches not to these meaning contents but to the site at which meaning contents present themselves, and Heidegger's question is not so much about how meaning can be determinate, but incomplete, but rather with how instances of Dasein can separate themselves, as *Selbstsein*, being a self, from the collective experiences of a historical location or situation, as fallen being with, *Mitsein*, and being with others, *Mitdasein*. Subsequent to *Being and Time*, the problem is how epochs in the sending and withdrawing of being are to be delimited with respect to each other.

In *Being and Time*, Heidegger challenges Husserl's notion of the living present as a stabilising, continuous domain of deployment of meanings, intentions and fulfilments, with his account of a centrifugal dispersion of time as vectors projecting into the past, into the present and the future, held together in the account of the temporal ecstasis of Dasein. Heidegger insists that of these three vectors, that associated with futurity is the most basic, because it is given determination in the understanding of mortality and finitude given in the structure of being-towards-death. The connection from this notion of ecstases to the tradition of Christian mystical experience poses a problem for the supposed neutrality of Heidegger's analyses. A resulting instability of this supposed ecstatic unity for Dasein, and for the notion of dividing time up into three dimensions, becomes clearer when the question is posed to its historical determination, as distinct in distinct epochs of the sending and withdrawal, forgetting and oblivion of being. This ecstatic temporality formally indicates an account of being as time, providing a horizon for thematising the various determinations of time as duration, interval, interruption, temporalisation and temporising, but Heidegger cannot satisfy himself that the formal indication can be fulfilled, in different ways in these different historical phases, and the enquiry of *Being and Time* is left incomplete. Heidegger's account of time is thus split between this incomplete earlier account of a precarious unity of Dasein's ecstatic temporality, giving provisional access to the temporality or temporising of being, and the mystery of how and why the sendings and withdrawals of being occur and have occurred in the ways registered in the tradition, in what Heidegger in the *Nietzsche* volumes of 1961, gives as a history of words for beings: *idea, energeia, actualitas, Wirklichkeit, Wille, Wille zur Macht.*[32]

How this history of being or of words for being is to be articulated on to the phases of thinking the time of Dasein and the phases of thinking the time of the event is subject to dispute. There are thus instabilities in the thinking of time both within each stage and between the analysis of Dasein in *Being and Time*, and the thinking of time in the artwork essay, 'The Origin of the Work of Art' (1936), roughly contemporaneous with the text, *Of the Event: Contribution to Philosophy* (GA 65: 1989). The papers on technology and thinking, published after the war, introduce the fourfold earth, sky, mortals and divinities, as a challenge to the apocalyptic time of heaven, hell, death and judgment. This appears to move the thinking of time into a phase where a separation between theological and philosophical concerns with time is more secure. The arrival of meaning in a clearing, in *Being and Time* called Dasein, articulated through the existentials, *Lichtung, Durchsichtigkeit* and *Rede*, and in these later writings reinscribed as the fourfold, retains the structure of phenomenological enquiry, in supposing that meaning is the result of processes of presentation in modes of givenness. These various accounts of time are challenged by Levinas, in favour of an overtly disjunctive thinking of time, in the arrival of the face of the other, which interrupts sameness and totality, invoking the arrival of the divine, in the face of the other. This requires a recognition of a more basic thinking of time and temporality as a trace of otherness, not susceptible to thematisation, as object of intending consciousness. This otherness is not susceptible to thematisation, and the temporality surmised to arise out of its trace cannot be made into the object of an intending consciousness. Thus Levinas appears to have departed from the circuit of phenomenology. However, his writings can also be seen to propose not a rejection but an inversion of the vectors of intentionality, with a movement from source of meaning, divinity, to its registration, taking the place of Husserl's move from thinking, to an object of thought.

Blanchot then contributes the thought that the line from thought content to thinking has a different configuration and tempo, from the line from thinking to thought content, thus introducing the notion of a curvature and swerve of thought, averted to in *Infinite Conversation* (1969).[33] This underlines the way in which these thinkers and texts variously provide distinct models for a line of time, quite distinct from the continuous line stretching from past to future or from future to past, hypothesised when time is thought of as an infinite series of 'now' points. This poses problems for each in responding to the others. Blanchot responds to Levinas in an essay from 1980, 'The Clandestine Companion', in the following way, by underlining the dispossession of identity in the modality of obsession:

> I would like to add an obsessional touch to these several notes. The book that Emmanuel Levinas has entitled *Otherwise than Being or Beyond Essence* is a philosophical work. It would be difficult not to take it as such, since philosophy, even if it concerns discontinuity and rupture, nonetheless solicits us philosophically. This book begins with a dedication, however, that I transcribe: '*To the memory of those who were closest among the six million assassinated by the National Socialists, and of the millions on millions of all confession and all nations, victims of the same hatred of the other, the same anti-semitism.*' How can one philosophize, how can one write within the memory of Auschwitz, of those who have said, oftentimes in notes buried near the crematoria: know what has happened, don't forget, and at the same time, you won't be able to.
>
> It is this thought that traverses, that bears the whole of Levinas' philosophy and that he proposes to us without saying it, beyond and before all obligation.
>
> (B: FF, p. 50)[34]

This hypothesis of an obsessional moment responds to and transforms Levinas' notion of a sense of responsibility and obligation taking possession of subjectivity, turning that subjectivity into the hostage for the safety of the other.

Levinas too responds to Derrida, in his contribution to the special edition of *L'Arc*, on Derrida's writings, in 1973, '*Tout autrement*', 'Wholly Otherwise', where there is to be found the following strong claim:

> May not Derrida's work cut into the development of Western thinking with a line of demarcation similar to that of Kantianism, which separated dogmatic philosophy from critical philosophy? Are we again at the end of a naivete, of an unsuspected dogmatism which slumbered at the base of that which we took for critical spirit? We may well ask ourselves. The Idea, as the completion of a series which begins in intuition without being able to end there; the Idea said to be 'in the Kantian sense of the term' would operate within intuition itself: a transcendental semblance itself generating metaphysics would create an illusion within presence itself, a presence that would ceaselessly be found to be wanting.
>
> (L: RL, p. 3)[35]

To this lapidary summation of a transformative relation between Derrida and Kant and an intertwining of the thinking of the Idea in the Kantian sense and the Kantian notion of transcendental illusion, as discussed in the previous chapter, this part of my enquiries seeks to add a disruption and displacement of Kant's notion of antinomy and of the critical notions of experience and of limits for deployment of that notion of experience.

In the later paper, 'Hostipitality', from the seminar of 1997, Derrida draws attention to an earlier version of this splitting of time, in Levinas' *Existence and Existents* (1948), between a thought of time as duration, or continuity, securing a continuous identity for subjectivity, and time as the interruptive instant, in which subjectivity is invoked, made into a problem, and constituted, in response to an arrival of alterity. The passage quoted by Derrida from Levinas' *Existence and Existents* runs as follows:

> Traditional philosophy, and Bergson and Heidegger too, remained with the conception of a time either taken to be purely exterior to the subject, a time-object, or taken to be entirely contained in the subject. But the subject in question was always a solitary subject. The ego all alone, the monad, already had a time. The renewal which time brings with it seemed to classical philosophy to be an event which it could account for by the monad, an event of negation. It is from the indetermination of nothingness, which the instant which negates itself at the approach of the new instant ends up in, that the subject was taken to draw its freedom. Classical philosophy left aside the freedom which consists in not negating oneself, but in having one's being pardoned by the very alterity of the other. It underestimated the alterity of the other in dialogue where the other frees us, because it believed there existed a silent dialogue of the soul with itself. In the end the problem of time is subordinate to the task of bringing out the specific terms with which dialogue has to be conceived.
>
> (L: EE, p. 94)[36]

Levinas' questioning of time is embedded in an analysis of the modes of prayer and meditation, scriptural interpretation and submission to clerical authority, through which access to the sources of pardon and salvation is held open. Derrida by contrast puts this circumscription of the questioning of time in question, by suspending any decision for or against the language of prayer and salvation.

It may, however, be false to claim that Heidegger is concerned only with the solitary ego, granted the concern in *Being and Time* with the problem of individuating Dasein and separating it from its collective conditions of existing as fallen 'they-self', and granted his unsolved problem with the determinacy of sendings of being. Nevertheless, this outline of a problem concerning time remains instructive. The implications of these invocations of a promise of salvation and a promise of pardon indicate the problem in relation to religious commitment and theological analysis, to be taken up in the next part of these enquiries. Remaining within the registers of human analysis, with no more than an invocation of a divine interruption, a reading of Blanchot permits both Levinas and Derrida to escape some of the paradoxes of seeking to write about structures, which they claim cannot be thematised. As indicated above, the challenge to the notion of the living present in Blanchot's writings comes from the notion of literature. In place of a notion of the living present as providing a thinking of the absolute flux of time, literature is the absolute condition for meaning, which is posed as the paradox of destruction and affirmation powerfully invoked in the image of the two slopes of writing. This takes place in the essay 'Literature and the Right to Death':

> If one looks at it in a certain way, literature has two slopes. One side of literature is turned toward the movement of negation by which things are separated for themselves and destroyed in order to be known, subjugated, communicated. Literature is not content to accept only fragmentary, successive results of this movement of negation: it wants to grasp the movement itself and it wants to comprehend the results in their totality. If negation is assumed to have gotten control of everything, then real things, taken one by one, all refer back to that unreal whole which they form together, to the world which is their meaning as a group, and this is the point of view that literature has adopted – it looks at things from the point of view of this still *imaginary* whole which they would *really* constitute if negation would be achieved.
>
> (B: WF, p. 330)

This provides an uncanny parallel to Husserl's description in *Ideas One* (1913) of the annihilation of what there is in the world, through the processes of reduction.[37] It is a notion of a literary absolute connecting back to that of early Romanticism, as discussed by Walter Benjamin, and in the analysis offered in *The Literary Absolute* by Jean-Luc Nancy and Philippe Lacoue-Labarthe.[38]

There is, however, also a strong link back to a reading of Hegel here, in the invocation of a movement of negation, which, by implication, is supposed to be only partially grasped by Hegel himself, and is to be thought instead in terms of this literary absolute. Blanchot thus effects a reversal of Hegel's claim to the superiority of philosophical discursivity, over that of literature and poetry. This inversion of Hegel then inflects Derrida's readings of Hegel, and underlines the significance for Blanchot of Heidegger's later affirmations of poetic thinking over philosophy. This is then supplemented by a description of the other side of literature:

> But there is another side to literature. Literature is a concern for the reality of things, for their unknown free and silent existence; literature is their innocence and their forbidden presence, it is the being which protests against revelation, it is the defiance of what does not want to take place outside. In this way it sympathises with darkness, with aimless passion, with lawless violence, with everything in the world that seems to perpetuate the refusal to come into the world.
>
> (B: WF, p. 330)

The importance of this as the moment of separation between Blanchot's thinking of death and that of Hegel should be remarked. It also sets out the distance between Derrida's concerns and those of an absolute idealism. Of more immediate concern here is the status of a conception of world, as given totality, which is put in question in Kant's Antinomies of Reason, and which is identified here by Blanchot as a denial of the existence of things. This is to be contrasted to the Husserlian notion of world, as that which remains when naturalised prejudices concerning what there is have been bracketed off. This suggests that the two slopes of literature cannot be conjoined, since their articulation requires the deployment of incompatible conceptions of world.

The paradox of this series of readings is that Blanchot, who is invoked by Levinas and Derrida, draws on some of the writings of Heidegger, and of Hegel, which are exactly put in question by Levinas and Derrida, in order to develop the position, on which Levinas and Derrida then draw. Blanchot also reads Heidegger to disrupt Hegel's subordination of responses to the artwork, first to an analysis of religion, and then a retrieval of philosophy. Blanchot affirms a privilege to the artwork over both religious and philosophical conception. Levinas affirms a privilege to a religious moment over philosophical thematisation, while Heidegger attempts to develop an alternative form of thematisation, to that proposed by Hegel, and indeed by Husserl. For Blanchot, Heidegger's later writings provide a certain inspiration concerning ways of thinking about an ineffability of language, summed up in Heidegger's gnomic phrase, '*Die Sprache spricht*', 'Language languages'.[39] For Levinas, there appears to be no decisive break or turn in Heidegger's philosophy, with the earlier affirmation of fundamental ontology, and the later affirmation of the God to come, taken as aspects of a single refusal of a divine source for thinking and responsibility.[40] For Derrida, there is a different continuity to be traced out, in Heidegger's attempts to excavate a thinking of time out of a response to a privileged Greek origin for philosophy. For this move both imposes a unity of Greekness on a diverse number of Greek thinkers, and excludes the possibility of heterogeneous, or multiple sets of origins for the thinking of time. Derrida distances himself from Heidegger's notion of a unitary givenness of such a retrieval, but it is not obvious that he is in any more sympathy with the Levinasian hypothesis of a divine origin. The contrast between Levinas' notion of a divine origin for meaning and Heidegger's notion of a lost Greek origin for meaning, comes to the fore in the differences between Heidegger's and Levinas' deployments of the notion of the trace.

The second of Derrida's *Geschlecht* essays, '*Geschlecht* II: Heidegger's Hand' (1985), ends with the following reflection:

> This primordiality is simply that starting from which things like metaphysics and Christianity are possible and thinkable. But what constitutes their arche-matinal origin and their ultra-Occidental horizon is nothing other that this hollow of a repetition, in the strongest and most unusual sense of the term. And the form of the 'logic' of this repetition is not only readable in this text on Trakl, but in everything that, since *Sein und Zeit*, analyses the structure of *Dasein*, the *Verfall*, the *Ruf*, care (*Sorge*) and regulates the relation

> of the 'most primordial' according to what is less so, notably Christianity. . . . Just as Heidegger requires a unique and gathering site for Trakl's *Gedicht*, he must presuppose that there is one single site, unique and univocal, for THE metaphysics and THE Christianity. But does that gathering take place? Has it a place, a unity of place?
>
> (D: 1985, G2 pp. 193–94)[41]

In order, with Heidegger, to set aside metaphysics in favour of ontology, or, with Levinas, to set aside ontology in favour of metaphysics, in order to set aside successive time in favour of interruption, or to set aside interruption in favour of successive time, they must first have been clearly distinguished, and it is the status of such distinctions which Derrida disputes, when he challenges Heidegger's and indeed Husserl's distinctions between derived and primordial time, in '*Ousia* and *gramme*: Note on a Note in *Being and Time*'.[42] Repetition and a retrieval of a single origin, and the pairing, derivation and primordiality, are for Derrida displaced in advance by the syncopations and detours articulated through his preferred disjunctive pairing of iteration and what remains, the *restance*.

A contrast between the thinking of Heidegger and of Levinas is sketched out by Derrida in the essay '*Différance*' in his invocations of their different concepts of the trace. Levinas' trace is of an immemorial time of divine origination. For the Heideggerian trace, Derrida reads Heidegger's essay, 'Anaximander's Saying' (1946), to which he then returns in *Specters of Marx*, to show that for Heidegger it is the trace of a Greek thinking, retained within but subordinated to a series of simplifications imposed by metaphysical delimitation.[43] This trace supplies a clue to another beginning, as a more originary thinking, which is identified by Heidegger as transmitted within metaphysics as its concealed double. Heidegger appeals for a revival of ontology, in the early writing, to achieve a return to these original Greek insights, and for an originary poetising, in the later writings, in order to arrive at modes of openness to givenness, which do not prejudge what is given in favour of a metaphysics of presence. In the essay, quoted by Derrida in '*Différance*', Heidegger writes:

> Only what is differentiated the present and presence (*das Anwesende und das Anwesen*) – becomes uncovered, but not insofar as it is differentiated. On the contrary the matinal trace of difference effaces itself from the moment that presence appears as a being-present (*das Anwesen wie ein Anwesendes erscheint*) and finds its derivation in a supreme (being) present (*in einem höchsten Anwesenden*).
>
> (H: OBT, p. 124)

Derrida comments on this:

> In this way the metaphysical text is *understood*; it is still readable and remains to be read. It proposes *both* the monument and the mirage of the trace,

> the trace as simultaneously traced and effaced, simultaneously alive and dead, alive as always to simulate in life its preserved inscription; it is a pyramid.
>
> Thus we think through without contradiction or at least without granting any pertinence to such contradiction what is perceptible and imperceptible about the trace.
>
> (D: 1968, SP pp.156–57)

And it is in conclusion to this text that Derrida invokes a certain Heideggerian hope, in the thought that it may all the same be possible to arrive at a 'proper word or unique name' for being, a hope which has no appeal for Derrida. Such naming stabilises the account to be given of the modes of givenness, and of temporalisation, of what there is, and of the arrival of entities as the entities they are, in a way which is precluded by the more disruptive mode of thinking the trace set out by Levinas in his essay, 'The Trace of the Other' (1963).[44]

This essay is implicitly invoked in the essay '*Différance*'. Derrida works towards Levinas' essay thus:

> With the alterity of the 'unconscious', we have to deal not with the horizons of modified presents, past or future – but with a 'past' that has never been nor will ever be present – whose 'future' will never be produced or reproduced in the form of presence. The concept of trace is therefore incommensurate with that of retention, that of the becoming-past of what had been present. The trace cannot be conceived – nor therefore can *différance* – on the basis of either the present or the presence of the present.
>
> (D: 1968, SP p. 152)

He makes a connection from Freud to Levinas, while seeking to mark their differences, thus:

> A past that has never been present: with this formula Emmanuel Levinas designates (in ways that are, to be sure, not those of psychoanalysis) the trace and the enigma of absolute alterity, that is the Other (*autrui*). At least within these limits and or this point of view, the thought of *différance* implies the whole critique of ontology undertaken by Levinas. And the concept of trace like that of *différance*, forms – across these different traces and through these differences between traces, as understood by Nietzsche, Freud and Levinas, (these 'authors' names' serve only as indications) – the network that sums up and permeates our 'epoch' as the de-limitation of ontology (of presence).
>
> (D: 1968, SP pp. 152–3)

Thus at this stage Derrida appears to situate the notion of *différance* closer to the Levinasian trace, of a past which has never been present, than to the

Heideggerian *Spur*, track, or path, for which a determinate topology may be hypothesised. However, in the course of his repeated engagements with the texts of Levinas and of Heidegger, he arrives at a position of greater neutrality between the thinking of Levinas and Heidegger, more akin to the stance of the earlier 'Violence and Metaphysics', insofar as he approaches a possible topology of the trace, and a tracing of topology. The next part of this section will continue the discussion of literature, death and time.

3.2 Literature, in the place of religion

> Literature now dispenses with the writer: it is no longer this inspiration at work, this negation asserting itself, this idea inscribed in the world as though it were the absolute perspective of the world in its totality. It is not beyond the world, but neither is it the world itself: it is the presence of things before the *world* exists, their perseverance after the world has disappeared, the stubbornness of what remains when everything vanishes and the dumbfoundedness of what appears when nothing exists. That is why it cannot be confused with consciousness, which illuminates things and makes decisions; it is *my* consciousness *without me*, the radiant passivity of mineral substances, the lucidity of the depths of torpor.
>
> (B: WF, p. 328)

The above citation is taken from Maurice Blanchot's essay 'Literature and the Right to Death' (1947), printed in his collection *La part du feu* (1949) (*The Work of Fire*).[1] The paragraph concludes: 'And it is not death either, because it manifests existence without being, existence which remains below existence, like an inexorable affirmation, without beginning or end – death as the impossibility of dying' (B: WF, p. 328). This sets up the possibility of the ghostlike living on, or *survivre*, which supervenes on actual lives:

> at the end of everything is fame; beyond, there is oblivion; farther beyond, anonymous survival as part of a dead culture; even farther beyond, perseverance in the eternity of the elements. Where is the end? Where is that death which is the hope of language? But language is *the life that endures death and maintains itself in it.*
>
> (B: WF, p. 336)

Derrida returns to this essay in a printed address closing the proceedings of what turned out to be the memorial conference for Blanchot in 2003. Under the title '*Maurice Blanchot est mort*', Derrida draws attention to the connection for Blanchot between the practice of literature, which is neutral with respect to death, and the political terror of the Jacobins, and by implication of the Stalinist Terror, where the political activists will their own death, if found to be in contravention of the aims of the political project.[2] In the course of this essay, Derrida coins a term '*litteraterreur*' (D: B, p. 606) to mark up a constant temptation to which writing can succumb, of installing an absolute which in politics implies a preference for death and glory, to life and compromise. Such a death and its justification in terms of a political programme reveals a challenge to the notion endorsed in Immanuel Kant's discussion of the illegitimacy of suicide in the *Groundwork for Metaphysics of Morals* (1785), of the duty to preserve one's life.[3] The shift from supposing that life belongs to God, to supposing that it attaches to a sovereign power, opens out into a discussion of political theology, and of jurisdictions over life and death.[4]

The distinction drawn by Blanchot between death and dying is one which echoes the distinction drawn by Heidegger in *Being and Time*, between the bio-physiological trajectory of dying, as a feature of a living organism, and his conception of being-towards-death, as an *apriori* existential determination of Dasein.[5] Blanchot's analyses mark up a lack of intersection between these two, since the being-towards-death of Dasein is absent precisely at the moment of the death of the biological organism. Heidegger's account is worked out in the opening sections of division two of *Being and Time*, as resulting from a reduction, in the style of Husserl, of everyday expressions and connotations concerning death to an essential meaning of an intimation, given to Dasein, of its own finitude as mortality. For Heidegger, this delimits Dasein as an individual and provides him with an account of a transition from indicating determinate existence as Dasein, to existing as a self, *Selbstsein*. It also provides him with a critique of the formalism of Kant's conception of a transcendental unity of apperception, and it sets up a fixed point in terms of which to develop the account of being, as the processes of temporalisation through which, amongst other determinations, a notion of time arrives. Derrida draws attention to the resonance between the two accounts given by Heidegger and by Blanchot concerning death, in the closing sections of *Aporias: Dying – Awaiting (One Another at) the 'Limits of Truth'*:

> When Blanchot constantly repeats – and it is a long complaint and not a triumph of life – the impossible dying, the impossibility, alas, of dying, he says at once the same thing and something completely different from Heidegger. It is just a question of knowing in which sense (in the sense of direction and trajectory) one reads the expression the possibility of impossibility.
>
> (D: 1993, AP p. 77)[6]

The awaiting/death of the title is a temporal determination, marked by Blanchot in this dying as an endless awaiting, which is interrupted by the arrival of an otherness, by death. As Derrida here marks up, there is also in relation to this analysis of possibility and impossibility, as with intentionality, a question of direction.

Derrida, unlike Levinas, is thus not inclined to read Blanchot's intervention as an inversion of Heidegger's claim concerning Dasein, possibility and death. The possibility of impossibility, which Blanchot converts into the impossibility of possibility, is evidently not the same as the possibility of the impossibility of Dasein, as determinate being, when not attested to. Heidegger thus writes of the cessation of Dasein, as determinate and as being:

> As potentiality-for-Being, Dasein cannot catch up with the possibility of death. Death is the possibility of the absolute (*schlechthinnigen*) impossibility of Dasein. Thus death reveals itself as that possibility which is most one's own, which is non-relational, and which cannot be caught up with (*unüberholbar*). (trans. mod)
>
> (H: SZ 250–51; MR, p. 294)

And for *schlechthinnigen*, it might be better to give 'straightforward' rather than 'absolute' as a translation. Derrida is, however, inclined to accept Levinas' thought that, by contrast to this primacy accorded to a relation of the everyday self to its authentic self, death is rather a response to the claim of the other, and, in mourning, is the affirmation of the continuing existence of the other in me. The work of mourning then marks for Derrida the impossible task of determinately delimiting Dasein as *Selbstsein*, from Dasein as *Mitsein*, being with, of delimiting Dasein as fully independent of all connection to others, living and dead.

In *The Writing of the Disaster* (1980), Blanchot explicitly invokes Heidegger's analysis of death, in a consideration of suicide:

> But what would the difference be between death by suicide and death by any other cause (if there is such a thing)? The difference is that the first, by entrusting itself to the dialectic (entirely founded upon the *possibility* of death, upon the use of death as power) is the obscure oracle which we do not decipher, but thanks to which we sense, and ceaselessly forget, that he who has been all the way to the end of the desire of death, invoking his right to death and exerting over himself a power of death, he who opens, as Heidegger said *the possibility of impossibility* – or again he who believes himself to be master of un-mastery – lets himself get caught in a sort of trap and halts eternally (halts, obviously, just an instant) at the point where, ceasing to be a subject, losing his stubborn liberty, and becoming other than himself, he comes up against death as that which does not happen or as that which reverses itself (betraying as though demented the mendacity of the dialectic by bringing it to its conclusion) – reverses the possibility of impossibility into *the impossibility of every possibility*.
>
> (B: WD, p. 70)[7]

The deployment of the language of dialectics, however, reveals that the interlocutor here is rather Hegel than Heidegger, since 'dialectics' is not a term used by Heidegger, and nor is it compatible with his analyses. The analysis runs together the analysis of death, in Hegel's terms, where a concept of desire opens out a scope for analysis, beyond that of finitude, and that of Heidegger, which deploys a mode of Husserlian reduction, to reveal in the everyday meanings attaching to death a pure essence of formally indicated meaning. Blanchot's description thus situates Heidegger's analysis of death, as one lying within the domain of human experience, as opposed to identifying its claim to constitute the very possibility of human identity. It thus fails to mark what is important for Heidegger: a distinction between an existentially understood transcendental death, through which meaning and order, and indeed an understanding of time, are to be constituted, and an empirical death, which is the end point, but not a fulfilment, of a process of living and dying, stretched out, as Heidegger puts it, between an actual birth and an actual death. It is in this way that Blanchot and Heidegger are obscurely in agreement about the differences between the process of dying, as empirical process, and a concept of death, as a term of philosophical analysis, concerning possibility and impossibility. Blanchot's phrase works in terms of a categorial modality; Heidegger's in terms of existential modes of Dasein.

The importance of a transformation of a notion of categorial possibility into one of an existential potentiality distinctive of Dasein is marked when Heidegger claims of being-towards-death:

> The closest closeness which one may have in being-towards-death as a possibility, is as far as possible from anything actual. The more unveiledly this possibility gets understood, the more purely does the understanding penetrate into it as the possibility of *the impossibility of any existence at all.* Death, as possibility gives Dasein nothing to be 'actualized', nothing which Dasein, as actual, could itself be. It is the possibility of the impossibility of every way of comporting oneself towards anything, of every way of existing. In the anticipation of this possibility it becomes 'greater and greater'; that is to say the possibility reveals itself to be such that it knows no measure at all, no more or less, but signifies the possibility of the measureless impossibility of existence. In accordance with its essence, this possibility offers no support for becoming intent on something, 'picturing' to oneself the actuality which is possible, and so forgetting its possibility. Being-towards-death, as anticipation of possibility, is what first *makes* this possibility *possible*, and sets it free as possibility.
>
> (H: SZ 262, MR, pp. 306–7)

This then is a limit as measureless measure, and it signals up a point of contact between Heidegger on Dasein's being towards death, and Kant, in the Third Critique, on the experience of the sublime.[8] For Heidegger, death reveals the possible impossibility of a determinate existence, rather than revealing the possibility of impossibility, without qualification. Even when Levinas, in the 1976 lectures *God, Death and Time* (1993) returns to these sections of *Being and Time* to give them a closer, more intensive reading than previously, this shift of scope of the notion of possibility is not remarked.[9] Thus when Levinas remarks that death is not so much the possibility of impossibility, but the impossibility of possibility, re-inscribing Blanchot's notion of an impossible death, he does not rehearse the Heideggerian phrase about death as the possibility of the straightforward impossibility of Dasein. For Levinas, the notion of Dasein has the status of an unproven hypothesis, by contrast to Heidegger's deployment of it to demarcate a possible horizon for enquiry. Thus this phrase of Blanchot's which may not be incompatible with the analyses of Heidegger is turned by Levinas into a challenge to Heidegger's analyses.

The importance for Heidegger of the constitution of possibility is reaffirmed in the account he gives of time and its temporalisation, both in *Being and Time* and in the summary of his views on time given in the 1928 lectures, *Metaphysical Foundations of Logic: Starting out from Leibniz* (GA 26, 1978).[10] Heidegger there clarifies the notion of ecstatic temporality, introduced in *Being and Time.* He relates it to the three basic vectors of Husserl's analyses of the relation between intentionality and time: expectancy, retention and making present, and to the distinction between the Greek term *ecstasis* and the Latin term, *raptus*, with its Christian eschatological overtones:

> The ecstasy mentioned here, stepping out of itself (*ekstasis*) is to some extent a *raptus* (rapture). This means Dasein does not become gradually expectant by traversing serially the beings that factually approach it as things in the future, but this traversing rather goes gradually through the open path made way by the raptus of temporality itself. Now this is true, in a corresponding manner of retention and making-present. And we therefore call these three basic phenomena the *ecstases* of temporality. Temporality is itself the self-unifying ecstatic unity in ecstatic temporalization.
>
> (H: MFL, p. 205)

Derrida's discussions of time in the writings of Husserl, Heidegger and Levinas puts in question any such unification of these ecstases; and raises a doubt about their capacity to coordinate the divergent strands of possible response to Husserl's opening out of the thought of a difference between time, and its modes of presentation, in temporalisation. Heidegger introduces his notion of time as contrastive to 'the common way of posing the question about time' (H: MFL, p. 197). He continues on the next page:

> All these descriptions of time, known both to the common as well as the philosophical understandings, cannot have been simply arbitrary fabrications and inventions. The essence of time must itself make these kinds of conceptions possible and even plausible. Yet none of them touches exactly the metaphysical essence of time.
>
> (H: MFL, p. 198)

The surprise here is that even after the incomplete publication of *Being and Time* in 1927, Heidegger is still writing and thinking in terms of an eidetic reduction of time to an essence.

He goes on to set out what he supposes that essence to consist in and how it is possible to arrive at this determination:

> 1: The essence of time has an ecstatic character.
> 2: Together with this ecstatic structure there is a horizonal character which belongs to time.
> 3: Time neither passes not remains but it temporalizes itself. Temporalization is the primal phenomenon of 'motion'.
> 4: Time is not relative to sensibility but is more primordial than sensibility and than mind and reason as well. Here we suppose, of course the only conception of reason with which we are acquainted, that of finite reason.
> 5: Methodologically we should note that, because it constitutes the metaphysical continuity of Dasein, time is not intelligible if Dasein is construed in some sort of theoretical schema, whether it be as a psychical whole, as cognitive-volitional subject, as self-awareness, or as the unity of body, soul and mind. Moreover, the analysis of Dasein must select for its guiding horizon the horizon which, in factical existence, continually guides Dasein's being-toward-itself, in its being-with, with others, and in its relation to beings unlike Dasein, prior to, outside of and despite all theory.
>
> (H: MFL, p. 198)

Thus Heidegger here affirms the priority of a concept of finite reason and the priority of the factual existence of Dasein as the site at which meaning and time present themselves, as opposed to any theoretical construction or thematisation. He also here makes clear that the fundamental ontology of the analytic of Dasein is provisional precisely because it is preparatory for the determination of such a metaphysics of Dasein. The analysis of *Being and Time* is marked by the double structure and double temporality of provisionality, that which Derrida invokes in terms of a logic of presupposition. This double structure, anticipating in a formal indication the determination to be arrived at, provides access to what, for Husserl, is a logic founding expectations of a fulfilment for formal indications of the meaning of being, and, for Heidegger, is an existential account of time, holding these formal indications in relation to each other, such that there might be a meaning of being. Heidegger thus displaces the Kantian unity of space and time in this doubled sense: the temporal flow of everyday living as human being and of theoretical construction in causality are brought to a standstill in the ontological questioning of how it is possible to have an understanding of being; and the unity of that humanity is referred back for its temporal unity in the temporality of Dasein as ecstatic.

In a preceding section of these lectures, Heidegger breaks off his analysis in order to set out a problem posed by certain responses to the conception of fundamental ontology of *Being and Time* and, implicitly, by Husserl's construal of the analytic of Dasein as nothing but a philosophical anthropology. Heidegger remarks:

> Fundamental ontology, as the analysis of the existence of Dasein, constitutes the approach to the problem. The analysis proceeds solely with the purpose of a fundamental ontology; the point of departure, execution, limit and mode of concretizing certain phenomena are governed by this purpose. The understanding-of-being is to be brought to light by way of Dasein's mode of being, which is primarily existence. The constitution of Dasein's being is such that the intrinsic possibility of the understanding of being which belongs essentially to Dasein becomes demonstrable. The issue therefore is neither one of anthropology nor of ethics, but of this being in its being as such, and thus one of preparatory analysis concerning it; the metaphysics of Dasein itself is not yet the central focus.
>
> (H: MFL, p. 136)

He then lays out ten guiding principles for an analysis of Dasein, and adds two further guiding statements: '11. This metaphysics of Dasein, first as an analysis, can be attempted only in the free projection of the being-constitution itself' (MFL, p. 139), and: '12. The ontological interpretation of Dasein's structures must be concrete with regard to the metaphysical neutrality and isolation of Dasein. Neutrality is in no way identical with the vagueness of a fuzzy concept of a "consciousness as such".' (MFL, p. 140). It is this notion of neutrality to which Derrida draws attention in his first paper entitled '*Geschlecht* I: Sexual Difference, Ontological Difference' (1983).[11]

In that essay, Derrida resumes the context in which Heidegger's remarks about the neutrality of Dasein occur and he focuses on the following guiding principle, number seven: '7: The transcendental dissemination proper to the metaphysical essence of neutral Dasein as the binding possibility of each factical existential dispersion and division, is based on a primordial feature of Dasein that of thrownness' (MFL, p. 138). Derrida draws attention to a slide from writing of a transcendental dissemination to an existential thrownness, and from dispersion, both transcendental and existential, to an existential multiplication of determinations. Derrida then adds a surprising and challenging observation:

> This order of implications opens up thinking to a sexual difference that would not yet be sexual duality, difference as dual. As we have already observed, what the course neutralised was less sexuality itself than the 'generic' mark of sexual difference, belonging to one of two sexes. Hence, in leading back to dispersion and multiplication, may one not begin to think a sexual difference (without negativity, let us clarify) not sealed by a two?
>
> (D: 1983, G1, BTB p. 401)

And he then adds 'The withdrawal of the dyad leads toward the other sexual difference. It may also prepare other questions. For instance, this one: how did difference get deposited in the two?' Thus Heidegger's insistence on neutrality leads to a questioning of the dual nature of sexual difference. Sexual difference starts out as one, becomes two, or three, and, thereby, acquires the form of a three, plus or minus one.

By contrast, when Blanchot writes of neutrality ('*le neutre*') there is another quite other effect. In the contribution to a failed dialogue with Jean-Luc Nancy, under the title '*Maurice Blanchot est mort*', Derrida concludes by citing sections on the neuter from Blanchot's book *The Step (not) Beyond/Le pas au-delà* (1973):

> If being reads itself, writes itself in the neuter, it is not, however, the case that the neuter comes before being, nor only that the neuter would give itself under the veil of the difference between being and beings, neither being nor beings (rather the beyond of the two or the hither of the between-the-two), but that the neuter averts it in gently dissuading it from any presence, even a negative one, neutralizing it to the point of preventing it from being called the being of the neuter, even while leading it into the infinite erosion of negative repetition.
>
> (B: SB, p. 76; D: 2003, B p. 622)

Derrida's essay concludes by citing the following fragment:

> The neuter, the gentle prohibition against dying, there where from threshold to threshold, eye without gaze, silence carries us into the proximity of the distant. Word still to be spoken beyond the living and the dead, *testifying to the absence of testimony.*
>
> (Ibid.)

And in French the word is '*attestation*', the *Bezeugung*, which Dasein gives to itself as that which in resoluteness takes on is own potentiality for being (H: SZ 267, MR, p. 311). Even more strikingly, this Neuter, which prohibits dying, takes on the role of Husserl's conception of transcendental life, as discussed by Derrida in *Speech and Phenomenon*, as beyond the distinction between death and life.

There is here a logic of seeking not to decide between the accounts of time offered by Levinas, Heidegger, Blanchot, and indeed Freud. In this Derrida is assisted by paying attention to incommensurabilities in the terms through which these accounts are articulated. The refusal to choose permits Derrida in *Aporias: Dying – Awaiting (One Another at) the 'Limits of Truth'* (1993) to move the questioning of time and death back from an apparent series of antinomies between the determinations of time offered by Heidegger as primordially finite, and Levinas' refusal of finite time, in the direction of a Greek notion of *aporia*. This notion of *aporia* is then confronted with responses to the traditions of biblical hermeneutics, presented on the one side by Heidegger's reading of Paul's *Epistles*, and the writings of Kierkegaard, and on the other, by the concealed transmission of Judaism in Marrano communities, and Levinas' Talmudic readings, which variously rehearse some of the paradoxes of faith, text and inherited traditions. For the revival and affirmation of a Greek notion of *aporia* has different implications for Heidegger and for Derrida, with Heidegger taking it up, in his notion of the forest paths which lead nowhere, the *Holzwege* of his 1950 collection of essays.[12] For Derrida the *aporia* becomes a question of the topology and the topography of the circumstance in which he finds himself: inheriting traditions to which he identifies himself as not belonging.

For Heidegger *aporia* indicates a possible affirmation of a thinking being, even if in order to do so it is necessary to invoke the thought that 'Only another God can save us now.'[13] For Derrida, the rethinking of the impasse of speculative thought entails its disruption by appeal to practical, historically given contexts of reasoning, not resolving paradox, but living it in the modes of their intensification, of the analyses of forgiveness, the promise, the gift. In *Aporias*, Derrida opens the analysis of *aporia* by considering: 'that the limits of truth are borders that must not be exceeded' (p. 1), in a clear invocation of the Kantian notion of the limits of the deployment of concepts in the pursuit of making knowledge claims. Instead, however, of a discussion of Kant, there is an invocation of the name of Diderot, and of Seneca, and a deflection of the line of questioning in the following way: 'What about borders with respect to death? And about borders of truth and borders of property? We are going to wander about in the neighbourhood of this question' (D: 1993, AP p. 3), and he points his themes thus:

> Thus, according to Seneca, there would be a property, a right of property to one's own life. In sum, the border (*finis*) of this property would be more essential, more originary, and more proper than those of any other territory in the world.
>
> (D: 1993, AP p. 3)

arriving at the thought 'that all these propositions whatever their modality, involve a certain pas (step, not). *Il y va d'un certain pas.* (It involves a certain step/not; it goes along at a certain pace)' (D: 1993, AP p. 6). This 'not' is the 'not' invoked by Blanchot, but it is also the 'not' of avoiding speaking explored in relation to negative theology, to be discussed in the next part of these enquiries. It is a pace or tempo quite as distinctive as the provisional projection, providing the looping tempo of Heidegger's enquiries, which, as the enquiry proceeds, reveals the ontological groundings out of which the apparently ungrounded phenomenal appearances arrive. For Derrida, the pace is one of an alternation between hesitation and precipitation, which has already been noted as a feature of his readings of Husserl.

Derrida alludes to his own earlier essay, '*Ousia* and *Gramme*: Note on Note from *Being and Time*', and remarks:

> The simple question from which I was trying to draw out the consequences (and from which one may never finish drawing them out) would be this: What if there was no other concept of time than the one that

> Heidegger calls 'vulgar'? What if, consequently, opposing another concept to the vulgar concept were itself impracticable, non-viable, impossible? What if it was the same for death, for a vulgar concept of death? What if the exoteric aporia therefore remained in a certain way irreducible, calling for an endurance, or shall we rather say an *experience* other than that consisting in opposing, from both sides of an indivisible line, an other concept, a non vulgar concept, to the so called vulgar concept?
>
> (D: 1993, AP p. 14)

Experience here is transposed into a notion of endurance, as temporally inflected. The intent of such a thinking of time and death must be not to oppose vulgar and non-vulgar concepts, nor indeed to oppose a concept of time as finite to a concept of time as infinite, for these would only be versions of each other. Thus the concern is with a certain 'not'. The intent of such a thinking of time and death must be to invent a form of thinking, as the experience of the impossible. Derrida continues:

> What would such an *experience* be? The word also means passage, traversal, endurance, and rite of passage, but can be a traversal without line and without indivisible border. Can it ever concern precisely (in all the domains where the question of decision and of responsibility that concern the border – ethics, law, politics, etc. – are posed), surpassing an aporia, crossing an oppositional line or *else* apprehending, enduring and putting in a different way, the experience of the aporia to a test? And is it an issue here of an either/or? Can one speak – and if so, in what sense – of an *experience of the aporia*? An experience of *the aporia as such.*
>
> (D: 1993, AP pp.14–15)

Or indeed, how is speaking of a certain experience to be avoided?

Derrida recalls a move made in the earlier *The Other Heading: Reflections on Today's Europe* (1991), drawing attention to the difference between hypothesising antinomy and encountering *aporia*:

> I suggested that a sort of nonpassive endurance of the *aporia* was the condition of responsibility and of decision. *Aporia* rather than antinomy: the word antinomy imposed itself up to a certain point, since in terms of the law (*nomos*) contradictions or antagonisms among equally imperative laws were at stake.

And he then declares:

> However, the antinomy here better deserves the name aporia insofar as it is neither an 'apparent or illusory' antinomy, nor a dialectizable contradiction in the Hegelian or Marxist sense, nor even a 'transcendental illusion in a dialectic of a Kantian type' but instead an interminable experience.
>
> (D: 1993, AP p. 16)

Having demarcated the move with respect to these possible misalignments, he then continues:

> Such an experience must remain such if one wants to think, to make come, or to let come any event of decision or of responsibility. The most general and therefore the most indeterminate form of this double and single duty is that a responsible decision must obey an 'it is necessary' that owes nothing, it must obey a *duty that owes nothing, that must owe nothing in order to be a duty*, a duty that has no debt to pay back, a duty without debt and therefore with duty.
>
> (D: 1993, AP p. 16)

This marks the shift from the privileged modality of necessity, in the First Critique to the privileged modality of duty, in the Second Critique, and intimates an inseparable connection between them. This implies that the objective necessities of the analysis of natural relations make no sense until that analysis is set back into an account of a recognition of a necessity, which requires an agent's affirmation, to bring it into effect. This remark thus moves across the registers of Kantian pure reason to a consideration of Kantian practical reason, and its affirmation of a concept of duty, a move licensed by the thought that in the *Critique of Pure Reason* there is a delimitation in terms of the time and space of human existence, whereas in the Second Critique the time horizon opens out to consider the immortality of the soul and a non-finitude of freedom, as set out within the spheres opened out in the infinities of a divine intending.[14] Such an affirmation of duty, with respect to divine law and creation is transposed by Heidegger into his secularised notion of attestation and testimony, *Bezeugung*, mentioned above.

Derrida then adduces:

> This formulation of the paradox and of the impossible therefore calls upon a figure that resembles a structure of temporality, as instantaneous dissociation from the present, a *différance* in being-with-itself of the present, of which I then gave some examples. These examples were not fortuitously political.
>
> (D: 1993, AP p. 17)

And he continues:

> In the end, the entire analysis concerned the very logic of exemplarism in any national or nationalist affirmation, particularly in Europe's relation to itself. In order to gain time, and before closing this backtracking that has the form of premises – forgive me, I needed to do so – I will rapidly mention the first seven *aporias* that concern the theme of this conference. Each of them puts to the test a *passage*, both an impossible and a necessary passage, and two apparently heterogeneous borders. The first type of border passes among *contents* (things, objects, referents, territories, countries, states,

> nations, cultures, languages, etc.), or between Europe and some non-Europe for example. The other type of borderly limit would pass between a *concept* (singularly that of duty) and an other, according to the bar of an oppositional logic.
>
> (D: 1993, AP pp. 17–18)

This specification of two types of border becomes three a few pages later: the first, between contents, a second, as a separation between distinct disciplinary discursivities, and then a third as the 'lines of separation, demarcation or opposition between conceptual determinations' (D: 1993, AP p. 23).

Now this very distinction between an oppositional logic of concepts and a series of independently given material contents is remarkable, as one which Husserl, for example, displaces by virtue of the parallelism between intentional relations between thought contents and material contents, and a relation of constitution between concepts and ontologically determinate entities. Derrida continues:

> The affirmation that announced itself through a negative form was therefore the necessity of *experience* itself, the experience of the aporia (and these two words that tell of the passage and the non-passage are thereby coupled in an aporetic fashion) as endurance or as passion, as interminable resistance or remainder.
>
> (D: 1993, AP p. 19)

Derrida remarks that the plural logic of the *aporia* takes shape, in the first place as non-passage arising for a simple closing of the border, '(exemplarily during war)' (AP p. 20). In the second place there is a non-passage when there is no longer a border to cross; no longer an opposition between two sides. He then adduces a third non-passage, where 'its elementary milieu does not allow for something that could be called passage, step, walk, gait, displacement, or replacement, a kinesis in general. There is no more path (*odos*, *methodos*, *Weg* or *Holzweg*)' (p. 20), with an obvious reference back to Heidegger's analyses of losing the path, in *Off the Beaten Track/Holzwege*. This has the following consequence:

> The coming or the future advent of the event would have no relation to the passage of what happens or comes to pass. In this case there would be an aporia because there is not even any space for an *aporia* determined as experience of the step or of the edge, crossing or not of some line, relation to some spatial figure of the limit. No more movement or trajectory, no more *trans* (transport, transposition, transgression, translation, and even transcendence). There would not even be any space for the aporia because of a lack of topographical conditions or more radically because of a lack of the topological condition itself.
>
> (D: 1993, AP p. 21)

Important, here, is the distinction between topographical conditions, and the topological, invoked as their transcendental condition. This relation will of course be reversed, with the transcendental shown to occur only in its topographically singular occasions. The insertion of the term 'transference' in this list would point a direction for further enquiry.

In the text which follows, Derrida short circuits a discussion of some of the major themes of Heidegger's existential analytic of death. He draws attention to the paragraph in section 49 of *Being and Time* where Heidegger sets out the claim of a priority of fundamental ontology over any speculative theology:

> Only when death is conceived in its full ontological essence can we have any methodological assurance in even *asking* what *may be after death*; only then can we do so with meaning and justification. Whether such a question is a possible *theoretical* question at all will not be decided here. The this-worldly ontological Interpretation of death takes precedence over any other-worldly speculation.
>
> (H: SZ 248, MR, p. 292)

It is the security of this distinction which the juxtaposition of the enquiries of Heidegger and Levinas concerning time, death, mortality and finitude compels Derrida to draw into question. Between Heidegger and Levinas, the questioning of death leads to a specification of a difference between a priority accorded by Heidegger and indeed Kierkegaard before him to the self-to-self relation of the call of conscience, and by Levinas and Buber, to the arrival of otherness and an affirmation of an otherness-in-me.

For Levinas, the thinking of death arrives always as a responsibility for the vulnerability of the other, whom I may already have condemned to death. This informs the urgency of the command: thou shalt not kill. Levinas' essay from 1984, 'Ethics as First Philosophy' puts the problem like this:

> From the beginning there is a face to face steadfast in its exposure to invisible death, to a mysterious forsakenness. Beyond the visibility of whatever is unveiled, and prior to any knowledge about death, mortality lies in the Other.
>
> Does not expression resemble more closely this extreme exposure than it does some supposed recourse to a code? True self-expression stresses the nakedness and defencelessness that encourages and directs the violence of the first crime: the goal of a murderous uprightness is especially well-suited to exposing or expressing the face. The first murderer probably does not realise the result of the blow he is about to deliver, but his violent design helps him to find the line with which death may give an air of unimpeachable rectitude to the face of the neighbour: the line is traced like the trajectory of the blow that is dealt and the arrow that kills.
>
> (L: LR, p. 83)[15]

This, then, is a quite other line of time which in the moment of murder misleadingly casts the victim in the role of innocence, rather than opening up the ontology of the human condition, as a separation from its divine origins. The respective guilt and innocence of victim and perpetrator is not the issue. It is the defiance of divine law which both reveals and conceals its priority.

The discussion continues, in a following paragraph:

> But in its expression, in its mortality, the face before me summons me, calls for me, begs for me as if the invisible death that must be faced by the Other, pure otherness, separated in some way, from any whole, were my business. It is as if that invisible death, ignored by the Other, whom already it concerns by the nakedness of its face, were already 'regarding' me, prior to confronting me, and becoming the death that stares me in the face.

The implications of this are summed up thus:

> The other one's death calls me into question, as if, by my possible future indifference, I had become the accomplice of the death to which the other, who cannot see it, is exposed; and as if, even before vowing myself to him, I had to answer for this death of the other, and to accompany the Other in his mortal solitude. The Other becomes my neighbour precisely through the way the face summons me, calls for me, begs for me, and in so doing recalls my responsibility and calls me into question.
>
> (L: LR, p. 83)

The two key elements in the argument are: 'A guiltless responsibility where by I am none the less open to an accusation of which no alibi, spatial or temporal could clear me', and: 'Responsibility for my neighbour dates from before my freedom in an immemorial past, an un-representable past that was never present and is more ancient than consciousness of . . . ' (L: LR, pp. 83–84). The tortuous syntax confirms the thought that indicative verbal determinations carry with them indications of possible future fulfilments of meaning, and with it the twin danger of supposing that there is a given time, to which responsibility may be connected back. This attempted disengagement from such structures reveals the movement whereby continuity and cumulative structure is disrupted for Levinas by the arrival of the infinite claim in the face of the other; and by the pulsation, within that instant of arrival, of a past which was never present, before any separation between God and what was created.

There can be no doubt that the analysis of Dasein, the analysis of determinate being, with its structure as stretched out between birth and death, as always being-toward-death, displaces the notion of the living present. It also displaces Husserl's account of intentionality, first in favour of the analysis of relatedness to ready-to-hand entities in a world, and then to the temporal ecstases, in which meaning arrives. What for Heidegger is missing from Husserl's phenomenology is a meditation on the arrival of a meaning of being, to provide the transition from doing phenomenology, to doing ontology. In the introduction to *Being and Time*, Heidegger writes: 'Being is *transcendens* itself' (SZ 38, MR, p. 62). It is that which permits the move from the supposed interiority of consciousness, as consciousness of a mental content, to consciousness as consciousness of determinate, consciousness-independent meanings. He goes on to specify what is distinctive about the transcendence of Dasein, of determinate being, which is 'that it implies the possibility and necessity of the most radical individuation'. This transcendence provides Dasein, as determinacy of existence, with its access to what there is in the world, and, when articulated as the givenness to Dasein of its temporality, this transcendence provides the structuring of the world, as a series of meaningful relations, through the disjoining and conjoining effects of temporalisation. This transcendence is then another name for intentionality, but is one through which the structure of intentionality as temporality comes to the fore. Indeed, Heidegger writes in a footnote much later in *Being and Time* that for him intentionality must be rethought as the structure of temporality:

> The thesis that all cognition has 'intuition' as its goal, has the temporal meaning that all cognising is making-present. Whether every science or even philosophical cognition aims at making-present need not be decided here.
>
> Husserl uses the expression 'make present' in characterizing sensory perception (Cf. his *Logische Untersuchungen*, first edition, 1901, vol. II pp. 588 and 620). This 'temporal' way of describing this phenomenon must have been suggested by the analysis of perception and intuition in general in terms of the idea of *intention*. That the intentionality of 'consciousness' is *grounded* in the ecstatical unity of Dasein, and how this is the case, will be shown in the following Division. [This Division has never been published. Tr.]
>
> (H: SZ 363, MR, p. 498)

For Heidegger an account of temporality includes within its scope an account of intentionality, delineating the modes of givenness of what there is, and embedding the account of intentionality, of protention and retention, within that broader conception of temporality as ontologically foundational.

Temporality is understood by Heidegger to be complemented not just with an account of historicality, as for Husserl, but also with an affirmation of the ordinary temporalisation of everydayness. This last is not to be confused with the ordinary conception of time, for which there are no such distinctions between the everyday, the temporal, as thematised, and the historical, as the condition of possibility for thematisation. This makes it possible to suppose that the ordinary concept of time provides a thematisation of time in which the pretence is made that thematising time makes no difference to the status of time. In the summer of 1928, in contemporaneous lectures, Heidegger surmises that the structure of inner time consciousness excavated by Husserl is the structure of time itself, thus suggesting that the ontological implications of the reduction of naturalised time to its pure form, given in transcendental consciousness, have not been fully drawn by Husserl. These are the lectures in which Derrida finds the discussion of Dasein, neutrality and, implicitly, sexual difference.

In these lectures, published as *The Metaphysical Foundations of Logic, Starting from Leibniz*, Heidegger comments on the first publication of Husserl's *Lectures on the Phenomenology of Inner Time Consciousness*, first given in 1904–05:

> These lectures are important, (aside from the concrete analyses of memory perception etc.) for the sharper development of intentionality beyond the *Logical Investigations*. That which Husserl still calls time consciousness, i.e. consciousness of time, is precisely, time itself in the primordial sense.
>
> We purposely call primordial time *temporality* in order to express the fact that time is not additionally on-hand, but that its essence is temporal. This means that time 'is' not, but rather temporalizes itself. Thus every attempt to fit time into any sort of being-concept must necessarily falter.
>
> (H: MFL, p. 204)

In this section twelve, Heidegger reverts to the account of time given in *Being and Time* as an ecstatic three-way movement outwards into expecting and awaiting, with respect to the future, having been, forgetting and remembering with respect to the past, and making present and overlooking, with respect to the present. These three ecstatic dimensions of the temporality of Dasein get levelled out in the mode of supposing that time simply passes into a single dimensionality with no alteration of tempo. It is thus from Heidegger that Derrida acquires his notion of the speeding up and the slowing down of temporal flows as important for any analysis of time.

In an essay from 1965, 'Intentionality and Sensation', Levinas quotes from Heidegger's brief introduction to the text of Husserl's 1904/05 lectures on time, where intentionality is marked up as a problem, not a completed explanation, and remarks: 'Heidegger's allusion can only urge us to investigate the original significance of intentionality in the way in which sensation is lived and in the dimension of time wherein it is lived.'[16] Levinas continues:

> The thread of time is a rectilinear multiplicity, a continuity of instants exterior to one another, without Bergsonian interpenetration. Inner time – the foundation of objective time and coextensive with it. Sensation that endures is spread out in this flow but if it is felt as an identifiable unity in this multiplicity of instants that exclude one another, it is because from each instant – thanks to an immanent and specific intentionality – the whole is retained in adumbration.
>
> (L: DE, pp. 141–42)

Levinas points out how for Husserl:

> Time does not arise out of an immobile eternity for a disengaged subject. Moreover, when in his descriptions of the constitution of time, Husserl uses

> expressions already having a temporal signification, he is not affirming a time behind time. One must recognise here a folding of time upon itself, a fundamental iteration.
>
> (L: DE, pp. 142–43)

Thus Levinas points out one way in which time and temporality for Husserl are doubled, reaffirming the doubling of the notion of intentionality, as both longitudinal and transverse movements, whereby the intending of the primordial intuition is stretched out across protention and retention.

Levinas concludes his reflections by radically transposing intentionality into a question of an ageing, with respect to the physiology which sustains sensation and the receptivity to impressions: 'The mystery of intentionality lies in the divergence ... or modification of the temporal flux. Consciousness is senescence and remembrance of things past' (L: DE, p. 145). This, however, leaves out another vital aspect of intentionality: its openness to disruption in the temporal determination of an *a-venir*, that which arrives out of the future and can be neither predicted, nor projected on the basis of a Husserlian protention, nor on any other extrapolation from given intuition or experience. The role this essay plays in the development of Derrida's thinking becomes clearer in the last paragraph, where Levinas writes:

> How can the *a priori* be an experience? Experience has always been understood as essentially uncertain in its claims and, in this sense, as leading thought astray. The novelty of phenomenology consists in reducing 'experience laying-claim-to-truth' to a conjuncture having signification by itself, that is to say, to the source of a transcendental work, starting from which the very notion of truth will only begin to have meaning. These significations are the original clarity. In Husserlian language, this turning around is called the Transcendental Reduction.
>
> (L: DE, p. 150)

For Husserl, there are more than the three dimensions, taken up in Heidegger's notion of ecstatic temporality. There is a fourth dimension, of natural time, and a fifth dimension, made up of the modes in which time arrives, otherwise known as temporalisation. He then goes on to reveal that this temporalisation is itself the result of its modes of presentation, inventing the notion of temporising, as the temporalisation which constitutes time itself, *Zeitigung*, alongside that of temporality, *Zeitlichkeit*, the form of its modes of givenness in consciousness. This is the moment of transition for Husserl from what is revealed in reduction, to making the claim that what is revealed in reduction is what objectively obtains. In the lectures contemporaneous to *Being and Time*, Heidegger attempts to develop his appropriation of Husserl's notions of temporality, *Zeitlichkeit*, and of temporising *Zeitigung*, but it not clear how this is to be conjoined to the account of the ecstases of Dasein. For the movement of *Zeitigung*, genesis, is one from beyond the limits of Dasein as determinate, finite, mortal being.

Where for Husserl and for Heidegger the source and origin of this genesis and temporalisation is left unspecified, and the registration of an arrival of the non-finite in finite consciousness is simply described as the task respectively of reason, as transcendental consciousness, or of an analysis of Dasein, Levinas complicates the structure by describing this as the arrival of an intimation of divinity as alterity, as a wholly other otherness, in the metaphysics of *Totality and Infinity: An Essay on Exteriority*, infinity as divinity, and the holy, beyond totality as the sphere of finite existence. As a consequence, intentionality acquires a double, ceasing to be solely the movement from consciousness to its objects, becoming as well, and first of all, a movement from a source of meaning and givenness, understood as divinity, towards finite consciousness. Thus any analysis of intentionality appears to be marked as incomplete and incompletable, for in so far as it is dependent on uninspectable and inconceivable activities of a divinity not amenable to conceptualisation, it defies appropriation in analysis and thematisation. These moves, however, reveal that Husserl's account of time and temporalisation is already marked by iteration and a scarcely retrievable otherness in the notions of genesis, history and the historical *aprioris*, marked up in the previous sections. These moves can also be followed up to reveal how for Husserl too there is not only a double movement of intentionality, as outlined in the *Lectures on Inner Time Consciousness*. There is the further doubling of an active synthesis, moving from formed consciousness to its objects, by underlying processes, in the movements of passive synthesis, leading from primary impressions to the formation of consciousness itself, replicating the movement from divine origin to human consciousness, but without the surmise or hypothesis of such a source.

Derrida's readings of each, Levinas and Heidegger, respect the specificity of their disruptive relations to Husserl, but refuse both moves in relation to Husserl, the move with Levinas of disambiguating Husserl's affirmation of a transcendental empirical doubling, in favour of a priority for a non-finite transcendental, and the move with Heidegger of disambiguating in favour of the determinacies of Dasein. Each, Levinas and Heidegger, modify their views in the course of their intellectual trajectories, but the contrast remains intact. Levinas, perhaps most significantly, takes into account the trenchant engagements with his views undertaken by Derrida in 'Violence and Metaphysics', which Levinas himself suggests leads to the change of tense and register of the subsequently published as *Otherwise than Being: Or beyond Essence*: its shifts, out of the indicative tense, the avoidance of unproblematic deployments of determinations of the verb 'to be', and the avoidance of main clause verbs. Heidegger shifts the focus for his account of the arrival of determinations of time and of being, from that of interrogating the conditions in which there occurs determinate existence, or Dasein, the central term of *Being and Time*, to the determinations of being, in epochal sendings, as *Ereignis*, the movement of a withdrawal of being in favour of conception. It might appear then that Derrida must thereby be committed to the stance of neither the one nor the other, but this is the stance which is analysed by Blanchot as the condition for writing. Derrida departs from the Blanchotian Neuter, and neither affirms nor denies an either/or or a neither/nor, between the accounts of time and the other, as disambiguations of Husserl, proposed by Heidegger and Levinas.

It is the refusal to choose between these three accounts of time, which positions Derrida's notion of time and impossible possibility, as neither that of Heidegger, nor that of Levinas, nor that of Blanchot; neither Dasein, nor *Ereignis*, nor infinity, as an intimation of the divine, nor writing as a death, revealing the impossibility of dying. The embedding of the various readings of Heidegger by Blanchot, of Blanchot by Derrida, of Derrida by Levinas, and of Husserl by all of them, generates a web of commentary and lines of exposition, which is duplicated by the implied conceptions of time and of temporality. The next part of these enquiries will consider whether it is possible for Derrida to pursue a thinking of time, which can be neutral between the commitments of literature and the problematics of writing, and the commitments of theology and the problematics of religion, by refusing the fixity of a demarcation between the two. The name 'Blanchot' has in this part been shown to mark the disruption of any exclusivity in a dialogue between Levinas and Heidegger. It prevents any stabilisation of their various enquiries into some series of opposed or antinomial views, on impossibility, death, otherness, or a priority of ethics over ontology, or of ontology over ethics.

Part IV

Religion without theology, theology without religion

4.1 Phenomenology as democracy to come

... in effect, this theology launches or carries negativity as the principle of auto-destruction in the heart of each thesis; in any case, this theology suspends every thesis, all belief, all *doxa* ...

–In which its *epokhe* has some affinity with the *skepsis* of scepticism as well as with the phenomenological reduction. And contrary to what we were saying a while ago, transcendental phenomenology as it passes through the suspension of all *doxa*, of every positing of existence, of every thesis, inhabits the same element as negative theology. One would be a good propaedeutic for the other.

(D: 1993, ON p. 66; D: 1992, DNT p. 308)[1]

There are various distinct strands of discussion in this dialogue, between two or more unnamed parties, printed under the ambiguous title *Sauf le nom: (post-scriptum)*, on saving and making an exception of the name, subtitle in brackets *post-scriptum*. Two issues immediately present themselves for attention: this delayed time of writing, in the post-script, and the erasure of the name, in the mode of plural anonymity. This mode of plural anonymity provides a link from Derrida's disruption of certainties about the name of God, in a multiplication of names, across languages and traditions, and his notion of a democracy to come, promising emancipation to the nameless, who are not yet born. My citation here suggests a third, more surprising issue: a surmised connection between negative theology and phenomenology, to which I shall return, by way of a long detour.[2] Phenomenology may be thought to have the form of negative theology, if not the name, since it advocates desisting from making statements about what there is, in order the better to reveal what there is. The preoccupation with the form of negative theology is also apparent in the idea of writing, as *post-scriptum*, as an after-effect of a moment of insight, which is suggested to be the form of writing of Augustine's *Confessions*, since their supposed addressee, God, may be presumed already to know what is confessed.[3]

This prompts one of the voices to remark:

This moment of writing is done for the 'afterwards' (*après*). But it also follows the conversion. It remains the trace of a present moment of the confession that would have no sense without such a conversion, without this

> address to the brother readers; as if the act of confession and of conversion having already taken place between God and him, being as it were written (it is an act in the sense of archive or memory) it was necessary to add a *post-scriptum* – the *Confessions*, nothing less – addressed to brothers, to those who are called to recognise themselves as the sons of God and brothers among themselves. Friendship here has to be interpreted as charity and as fraternity.
>
> (D: 1993, ON pp. 39–40)

This structure of time as postponement reveals what has happened, the conversion to God, only after the fact. Confession is the *après-coup*, the delayed affect, of conversion, and Augustine's text, the *Confessions* is understood to be addressed to its readers, as brothers in Christ, who are thus encouraged to recognise themselves in the condition of the one who may turn towards God. These readers are here addressed as brothers and sons, and Derrida quite deliberately underlines a sexism at the heart of this moment of Christian affirmation. He finds fraternity at once affirmed and denied, in the transmission of Christianity, and in the various affirmations of political ideals, which he analyses in *Politics of Friendship* (1994).[4] He thus also reveals a certain continuity and complicity between the religious and the political. The ambiguous status of this brotherhood, as both affirmed and denied, may too explain the hysteria displayed by Church authorities when confronted with the homosexuality of the other, and, indeed, of their own communities. This section of my enquiries is concerned with such continuities and hidden effects, which leave an imprint on the analysis of time, held in place between these various strands of mutuality and continuity. If these effects of delay and displacement are characteristic of time in general, but tend to be denied in thinking about everyday experiences of time, or indeed a natural scientific construction of time, it is plausible to suppose that religion and theology are designed to give expression to structures and experiences of time which are otherwise repressed, or obscured.

There are for Derrida a series of overlapping concerns, shared by political analysis, by an aporetics of literary activity, as analysed by Blanchot, and by the various inheritances of religion and theology, which are held in tension by his readings, in parallel, of some of the principal texts of these traditions, through the 1980s and 1990s. Some of his texts explicitly combine considerations, from these various strands of enquiry, with this further suggestion of a connection between negative theology, and a politics attributable to Husserl's phenomenological programme. This takes the form of a political theology, to which Derrida seeks to object, most explicitly in *Rogues: Two Essays on Reason* (2003) but implicitly already in earlier texts, when the analysis of a politics in Husserl's phenomenology comes to the fore in *The Other Heading: Reflections on Today's Europe* (1991) and in *The Gift of Death* (1992).[5] An objection to this political theology serves as a barrier between Derrida and any affirmation of Benjamin's inversion of political theology, into a politicisation of theological concepts. Derrida is convinced of the success of neither the inversion of the relation between politics and religion, attempted by Marx, in which religion is a mask for political oppression, nor the inversion of the relation between political and theological concepts, attempted by Benjamin, since such an inversion appears to leave the structural dominance of the pairing sacred/secular, divine/mortal intact. A diagnosis of a proximity and distance between Derrida and Benjamin is the end point of this section.

Derrida's more explicitly political texts are *The Other Heading: Reflections on Today's Europe* (1991), *Specters of Marx: The State of the Debt, the Work of Mourning and the New International* (1993), *Politics of Friendship* (1994) from the 1988–89 seminar, and *Voyous: Rogues: Two Essays on Reason* (2003), which might be thought to fill out the discussion of the New International, left underdeveloped in *Specters*.[6] I shall here insert a reading of some of the moves made in *Specters* into the context of some of the questions raised and not raised in the text *On the Name*, in which *Sauf le Nom* was translated into English. The spectres of *Specters of Marx* (1993) are those evoked by Marx himself, the spectre of communism haunting Europe and the spectral animation of commodities, as well as the shadows of religion and of a critique of religion which inform Marx's critique of political economy. The shadow of literature also haunts Marx's texts: he himself was often inclined to cite Shakespeare.[7] Perhaps for this reason Derrida cites the staging of the appearances of the ghost of *Hamlet, Prince of Denmark*, both to disrupt and to focus his account of the incompleteness structural to political critique and to the revolution announced as a historical necessity by Marx and Marxism.

In the course of the text Derrida makes connections between the notion of the *a-venir*, derived from the reading of Blanchot and Levinas, and an analysis of a concept of democracy, as democracy to come, or democracy as promised. This promise is understood on the model of the promise to Abraham, marked by the substitute sacrifice of the lamb for the son, of the messianic promise of redemption, and of the promise made to the apostles that Jesus would come again. Derrida imports this complicated temporality of the event which, in so far as there is faith, has already arrived, to explain the curious nature of democracy. This promise of democracy has, as analysed by Len Lawlor, this complex temporality of contingent historical conjuncture, categorical injunction and an afterlife in an indispensable double affirmation, in human communities of remembrance.[8] The notion of originary mourning, mooted in *Aporias*, is here taken up and recast as the unavailability of a common time of shared temporal durations. There is a collective experience of the lack of simultaneity experienced at an individual level, where friends necessarily outlive one another, installing as unavoidable a work of mourning for a lost commonality.

Towards the end of the second section, 'conjuring–marxism', Derrida inserts an aside to the following effect:

> Permit me to recall that a certain deconstructive procedure, at least the one in which I thought I had to engage, consisted from the outset in putting into question the onto-theo – but also archeo-teleological concept of history – in Hegel, Marx or even in the epochal thinking of Heidegger. Not in order to

> oppose it with an end of history of an anhistoricity, but, on the contrary, in order to show that this onto-theo-archeo-teleology locks up, neutralizes, and finally cancels historicity.
>
> (D: 1993, SM pp. 74–75)

The reading of Marx offered in *Specters* is thus contentious, not least because it brings that reading into close proximity with a series of philosophical concerns emergent from Derrida's ongoing encounters with Kant and Hegel, with Husserl and Heidegger. Onto-theology is the theoretical formation which Heidegger seeks to displace but finds himself, according to Derrida, necessarily reinstalling; archeo-teleological thinking is articulated by Hegel in terms of an emergence from and a return to a divine origin and goal for creation, and this structure is similarly both challenged by and reinstalled in Marx's attribution of a necessity to the overcoming of capitalist appropriations and distortions of the ideals of emancipation. Derrida continues the analysis:

> It was then a matter of thinking another historicity – not a new history or still less a 'new historicism' but another opening of event-ness as historicity that permitted one not to renounce, but on the contrary to open up access to an affirmative thinking of the messianic and emancipatory promise as promise: as *promise* and not as onto-theological or teleo-eschatological program or design. Not only must one not renounce the emancipatory desire, it is necessary to insist on it more than ever, it seems, and insist on it, moreover, as the very indestructibility of the 'it is necessary'. This is the condition of a re-politicization, perhaps of another concept of the political.
>
> (D: 1993, SM p. 75)

Thus Derrida disruptively deploys Heidegger's concept of the event, and of another beginning, to detach Marxist accounts of history both from new historicism and from the appearance, and indeed the actuality of imposing a freedom on human beings, within a pre-determined narrative of emancipation.

This other concept of the political opens out the distinction in French between *la politique*, politics as a structural condition, and *le politique*, the immediate strategies and forms of contestation.[9] The latter is what commonly passes for politics, the former is to be developed, in part borrowing from Heidegger's analysis of originary polemics and contestation, as outlined in Derrida's reading of Heidegger in the slightly earlier paper 'Heidegger's Ear: Of *Philopolemology* (*Geschlecht* IV)' (1989), to which Derrida refers back in a footnote towards the end of *The Gift of Death*. This structural political moment stands in a relation to political intrigue as the notion of historicity stands to history: the structural moment pertains even in the absence of affirmation of and engagement in the practice in question. In this sense they are transcendental and *apriori*, and play a role in determining the spatio-temporal conditions distinctive of the human condition. Derrida's analyses imply that the emancipatory promise is always conditional on the availability of technical means for enunciation, and for mitigating the effects of the human conditions, as animal, and as finite. This connects these analyses to the account of technology developed by Heidegger as an analysis of the dominant mode in which what there is comes to appearance in the current epoch.

A moment of materiality is articulated in the techniques for responding to human physiological vulnerability and finitude, as prone to illness, ageing, and all the complications of reproduction, and in the structures and systems of communication, information retrieval, and expression available at any historical juncture. Processes of analysis and human interaction are reliant on available linguistic resources, techniques of intervention, and organisation of archives of information and memory, revealing what is analysed by Bernard Stiegler as an originary technicity.[10] This presents an account of a prosthetic moment, occurring in advance of any formation of a thesis, a system of techniques making possible the determination of an embodied actuality, the *Verkörperung* of thought which Husserl supposes is given in fulfilled intuition. This analysis places a technical, prosthetic moment, supplementing an inadequately equipped natural organism, in advance of the formation and registration of that supposedly natural, organic growth, placing *techne*, techniques, or learned skills, in advance of *phusis*, or natural endowments.[11] This inversion places a time of technical registration of effects in advance of any supposedly natural series of such effects.

This inversion of priority stalls the development of any single account, and of any self-contained trajectory from a unique origin to a determinate outcome, in a teleology, revealing that there is no single trajectory, and no such singleness of origin. The plurality of endings marked in originary mourning is thus mirrored by a plurality of starting points, in originary technicity. This multiplication of starting points and the necessity of negotiating differences of register between invoking an originary mourning and invoking an originary technicity, opens up for redetermination the Kantian notion of transcendental aesthetics as the framing of space and time within which human experience takes place, for there is a contestation of the renunciation of analysis of beginnings and ends announced by Kant in the second Antinomy of Pure Reason. The contours of this rethinking are what is in contention between Derrida and Benjamin.

In *Specters*, Derrida responds to Fukuyama's text, *The End of History and the Last Man* (1992)[12]. He identifies an unhingeing of time, in the relation between capitalism, communism and Marxist critique, and he stages this unhinging by tracking back and forth between a reading of Marx and Engels's *The German Ideology* (1846) and a reading of Shakespeare's *Hamlet, Prince of Denmark*, twisting together considerations from a discussion of the ends of politics, a response to Husserl's phenomenology, and a legacy of faith.[13] At the opening of the text, Derrida gives a reading of an essay by Maurice Blanchot, 'Marx's Three Voices'.[14] This he positions between the controversial invocation of Heidegger's 'Anaximander Fragment',[15] and young Hamlet's protest in Shakespeare's play:

> The time is out of joint; o, cursed spite
> That ever I was born to set it right.

This 'out of joint' is brought into conjunction with the questions of injunction and apportioning, invoked by Heidegger in his translation of *dike* and *adikia* from the Greek of Anaximander, and as analysed in his essay 'The Anaximander Fragment' (1946). Blanchot's essay closes with the following paragraph, part of which Derrida quotes:

> The word *science* becomes a key word again. Let us admit it. But let us remember that if there are sciences, there is not yet science, because the scientificity of science still remains dependent on ideology, an ideology that is today irreducible by any particular science, even a human science; and on the other hand, let us remember that no writer, even Marxist, could return to writing as to a knowledge, for literature (the demand to write when it takes control of all the forces and forms of dissolution, of transformation) becomes a science only by the same movement that leads science to become in its turn literature, inscribed discourse, which falls, as always, within 'the senseless play of writing'.
>
> (B: F, p. 100)

Derrida oddly interrupts this before the reversal of direction of analysis towards a questioning of a dependence of analysis on its literary form.

For Marx, the difference between economic ideology and his own scientific political economy is key and, in a different register, this distinction is also key for Husserl. The earlier task of seeking out an essence of scientificity, in 'Philosophy as Rigorous Science' (1911), to ground the clamis for phenomenology over other forms of philosophy, gives way to the broader task, in *The Crisis of the European Sciences* (1938).[16] There Husserl discusses the emergence in history of a concept of science, which has the form of an *apriori*. This analysis of the genesis of the concepts of science and of apriorism is then to play a role in affirming, and restoring the claims to scientific status, and to scientific progress made on behalf of European science, and as a confirmation of European achievements, as a response to and challenge to the irrationalism with respect to science in evidence in the contemporary manoeuvres of the Nazis with respect to science. Derrida quotes Blanchot's paragraph just after remarking:

> Once again, here as elsewhere, wherever deconstruction is at stake, it would be a matter of linking an *affirmation* (in particular a political one), if there is any, to the experience of the impossible, which can only be a radical experience of the *perhaps*.
>
> (D: 1993, SM p. 35)

Derrida here is keener to emphasise this modality of uncertainty, rather than pursue a connection from Blanchot to Husserl, although he later makes decisive appeal to Husserl's analyses, and clarifies his notion of the 'to-come' by contrast to a notion of the living present.

Derrida implicitly invokes Bataille's notion developed in *Inner Experience*, as analysed in the earlier essay 'From Restricted to General Economy' (1968).[17] This explores the paradox that sovereignty cannot be deemed to be subject even to its own legislative authority:

> *Violence* of the law before the law and before meaning, violence that interrupts time, disarticulates it, dislodges it, displaces it out of its natural lodging: 'out of joint'. It is there that differ*a*nce, if it remain irreducible, irreducibly required by the spacing of any promise and by the future-to-come that comes to open it, does not mean only (as some people have too often believed and so naively) deferral, lateness, delay, postponement. In the incoercible differ*a*nce the here-now unfurls.
>
> (D: 1993, SM p. 31)

This 'here and now' clearly is to be distinguished from the 'now' of punctual moments arranged in the series, sequences and successions of linear time. The question is: how? Derrida continues:

> Without lateness, without delay, but without presence, it is the precipitation of an absolute singularity, singular because differing, precisely and always other, binding itself necessarily to the form of the instant, in *imminence and in urgency*: even if it moves toward what remains to come, there is the pledge [*gage*] (promise, engagement, injunction and response to the injunction, and so forth). The pledge is given here and, now even before, perhaps, a decision confirms it. It thus responds without delay to the demand of justice. The latter by definition is impatient, uncompromising and unconditional.
>
> (Ibid.)

He draws these remarks to a summary close with the enigmatic claim: 'No differ*a*nce without alterity, no alterity without singularity, no singularity without here-now' (p. 31). He thus links and disjoins '*différance*', neither a word nor a concept, to Levinas' notion of alterity, and then to a concept of singularity, thus disrupting and challenging Heidegger's 'Da', of Dasein, the singularity which rightly understood affirms itself as a self.

Derrida here affirms a distinction between an eschatological temporality of interruption and limit, and a teleological temporality, of duration and completion, showing how political events, as depicted in Shakespeare's play at any rate, depict a non-teleological breaking-open of time. The arrival of the ghost and Hamlet's response to it cannot be predicted or indeed ordered into time. Thus this 'here-now' breaks open any enclosure and subordination of the eschatological limit case to a teleology providing an overall trajectory of meaning. Writing of the future to come, Derrida claims:

> We must discern here between eschatology and teleology, even if the stakes of such a difference risk constantly being effaced in the most fragile and

> slight insubstantiality – and will in a certain way always and necessarily be deprived of any insurance against this risk. Is there not a messianic extremity, as *eskhaton*, whose ultimate event (immediate rupture, unheard-of interruption, untimeliness of the infinite surprise, hetereogeneity without accomplishment) can exceed, *at each moment*, the final term of a *phusis*, such as work, the production, as the *telos* of any history?
>
> (D: 1993, SM p. 37)

Thus there is a temporality of technicity, and of an originary prosthetics which escapes the closures of a temporality of *phusis*. This mobilises the temporality of Levinas' eschatology against the cumulative trajectory of history as imagined by Hegel, or indeed by Marx. This then was enough to upset a number of more orthodox readers of Marx, for Derrida uncouples revolutionary moments from any overarching account of a necessary success for those who make the revolution. Derrida's reading of Marx is thus quite distinct from those readings subordinated to securing a role for a party or a party leadership in relation to such necessary success, rather as his reading of negative theology does not affirm the infallibility of any pope, or privilege any college of theologians.

The account of the future, leading into the formulation of this distinction between eschatology and teleology, runs as follows:

> The opening must preserve this heterogeneity as the only chance of an affirmed or rather reaffirmed future. It is the future itself, it comes from there. The future is its memory. In the experience of the end, in its insistent, instant, always imminently eschatological coming, at the extremity of the extreme today, there would thus be announced the future of what comes.
>
> (D: 1993, SM p. 37)

And in this language of the extremity of the extreme there lies the echo of Bataille's disruption of the legal concept of sovereignty, by the hypothesis of an inner experience of sovereignty which always goes beyond and is irreducible to any legal system, or indeed to analysis in terms of a concept of justice. It is this excess of spontaneity, possibility and of the modality of the 'perhaps', over any given order, which Derrida indicates by resort to his reading of Heidegger's essay 'The Anaximander Fragment', which links the analysis in *Specters* to that of *Geschlecht IV.* Derrida's analysis continues:

> More than ever, for the future-to-come can announce itself as such and in its purity only on the basis of a *past end*: beyond, *if that's possible*, the last extremity. If that's possible, *if there is any* future, but how can one suspend such a question or deprive oneself of such a reserve without *concluding in advance*, without reducing in advance both the future and its chance? Without totalising in advance?
>
> (D: 1993, SM p. 37)

The aim is to release the future of what comes from any pre-determination, in advance, such that it might arrive out of future possibilities, not set in place by the possibilities revealed in the past, and thus exceeding the limits set out in the past, rather than realising an outcome already delineated in that past.

It is in this context that Derrida reads into Marx's writings, and into Marxist analysis, the thought of an experience of the emancipatory promise, which he grounds in the notion of the messianic without messianism. The linkage from Marx to the religious formations he analyses as ideological is explicitly made:

> we will not claim that this messianic eschatology common both to the religions it criticizes and to the Marxist critique must be simply deconstructed. While it is common to both of them, with the exception of the content [but none of them can accept, of course this *epokhe* of the content, whereas we hold it here to be essential to the messianic in general, as thinking of the other and of the event to come] it is also the case that its formal structure of promise exceeds them or precedes them. Well, what remains irreducible to any deconstruction, what remains as undeconstructable as the possibility itself of deconstruction is, perhaps, a certain experience of the emancipatory promise; it is perhaps even the formality of a structural messianism, a messianism without religion, even a messianic without messianism, an idea of justice – which we distinguish from law or right and even from human rights – and an idea of democracy – which we distinguish for its current concept . . .
>
> (D: 1993, SM p. 59)

And at this point Derrida refers back to the earlier paper, 'The Force of Law and the "Mystical Foundation" of its Authority' (1989), to which I shall return in conclusion to this discussion of religion and theology. This 'formal structure of a promise', and the 'formality' attributed to a 'structural messianism' recalls the analysis of formalism and formality in Husserl's account of meaning, as indeed does this invocation of an '*epokhe* of the content'.

At stake here, Derrida writes, is 'the very concept of democracy, as concept of a promise', and he continues:

> That is why we always propose to speak of a democracy *to come*, not of a *future* democracy in the future present, not even of a regulating idea, in the Kantian sense, or of a utopia – at least to the extent that their inaccessibility would still retain the temporal form of a *future present*, of a future modality of *living present*.
>
> (D: 1993, SM pp. 64–65)

The paradoxical nature of this analysis appears clearly in the following remark, which connects to the analyses of impossibility in *Aporias*:

> It would be easy, too easy to show that such hospitality without reserve, which is nevertheless the condition of the event and thus of history (nothing

> and no one would arrive otherwise, a hypothesis that one can never exclude of course) is the impossible itself, and that this *condition of possibility* of the event is also its *condition of impossibility*, like this strange concept of messianism without content, of the messianic without messianism, that guides us here like the blind.
>
> (D: 1993, SM p. 65)

The mention here of the blind is by no means fortuitous, for Derrida is about to advance an account of Husserl's phenomenology which no longer turns on an indefensible contrast between the visible and the invisible, as the metaphorics for modelling the apparent and the inapparent. In an extended footnote Derrida explicitly invokes Husserl's analyses, to show that for Husserl even a contrast between the real and the unreal does not privilege the real over the unreal, for he invokes a third term, the real determinations of essential structures, the *reell*, which reveals the derivation of what appears to be real or unreal from these previously determined real (*reell*) distinctions.

In section 5, '*Apparition of the inapparent:* the phenomenological conjuring trick', the implied relation to the phenomenology of Husserl comes to the fore. In a discussion of Marx's critique of Max Stirner's account of appearance and property, Derrida invokes a question of phenomenology first in relation to Marx's take up of Hegel's *Phenomenology of Spirit*, in which spirit presents itself in a series of increasingly adequate presentations. Derrida then invokes the phenomenology of Husserl, which does not inscribe the series of appearances to human beings within an account of an over-arching divine intending. This makes the converse claim that it is not the organisation of concepts which drives the move to greater adequacy, but the efforts of human intelligence, taking previous enquiry, affirming its goal of establishing truth, and improving on it. Derrida writes:

> We underscore here the objections that one might be tempted to address to the phenomenological principle in general. Two conclusions then: 1) the phenomenal form of the world itself is spectral; 2) the phenomenological ego (Me, You, and so forth) is a spectre. The *phainesthai* itself (before its determination as phenomenon or phantasm, thus as phantom) is the very possibility of the spectre, it brings death, it gives death, it works at mourning.
>
> (D: 1993, SM p. 135)

The phenomenal form of the world is an effect of the work of transcendental subjectivity, and individual identities are the effect of interaction and of a moment of retrospective constitution in the after-effects of living, dying and a writing of selves into determinacy, in the *mémoires*, memories and work of constitution that is given by Derrida this title, originary mourning.

Derrida identifies in the dispute between Stirner and Marx, an analysis of self-preservation, in which an abstract moment of analysis, a concept of the ego, or of identity, makes use of aspects of structures other than itself to preserve itself:

> To protect its life, to constitute itself as unique living ego, to relate, as the same to itself, it is necessarily led to welcome the other within (so many figures of death: differ*a*nce of the technical apparatus, iterability, non-uniqueness, prosthesis, synthetic image, simulacrum, all of which begins with language, before language). It must therefore take the immune defenses apparently meant for the non-ego, the enemy, the opposite, the adversary, and direct them at once *for itself and against itself.*
>
> (D: 1993, SM p. 141)

The connection from Marx back to the religions of the Book is made through a contrast to the thinking of futurity to be found in Husserl's phenomenology:

> The messianic, including its revolutionary forms (and the messianic is always revolutionary, it has to be), would be urgency, imminence but, irreducible paradox, a waiting without horizon of expectation
>
> (D: 1993, SM p. 168)

This 'waiting without horizon of expectation' marks a break with the conception of protention, projecting future meaning, advanced by Husserl, which is a waiting for satisfaction or disappointment of meaning intention, delimited through a conception of the horizon of all horizons, of all possible protending of meaning. Derrida continues:

> One may always take the quasi-atheistic dryness of the messianic to be the condition of the religions of the Book, a desert that was not even theirs (but the earth is always borrowed, on loan from God, it is never possessed by the occupier, says precisely the Old Testament whose injunction one would also have to hear); one may always recognise there the arid soil in which grew, and passed away, the living figures of all the messiahs, whether they were announced, recognised or still awaited. One may also consider this compulsive growth, and the furtiveness of this passage, to be the only events on the basis of which we approach and first of all name the messianic in general, that other ghost which we cannot and ought not do without.
>
> (D: 1993, SM p. 168)

And here the notion of hospitality is to be tied to the instance of welcome necessary to allow messiah to arrive, opening up a futurity unconstrained by horizonal expectation.

For the arrival of objectivity, there must be a subjective instance permitting that arrival while that subjective instance itself is spectral in relation to a notion of an objectively given world. In an extended footnote Derrida extends the analysis to include a response to Husserl:

> Of course, the narrow and strict conception of the phantom or the *phantasma* will never be reduced to the generality of the *phainesthai*. Concerned

> with the original experience of haunting, a phenomenology of the spectral ought, according to good Husserlian logic, to isolate the very determined and relatively derived field within a regional discipline (for example, a phenomenology of the image and so forth). Without contesting here the legitimacy or even the fertility of such a delimitation, we are merely suggesting the following, without being able here to go any further; the radical possibility of all spectrality should be sought in the direction that Husserl identifies, in such a surprising but forceful way, as an intentional but non-real (*non-reelle*) component of the phenomenological lived experience, namely the *noema*.
>
> (D: 1993, SM p. 189, fn 6)

The *noema*, be it remembered, is the term introduced in *Ideas One* for the thought content of the thinking, or *noesis*, from which the contents of transcendental consciousness are made up. This pairing in Husserl's analyses displaces any pairing between an Aristotelian *nous*, as non-finite intellection, and its objects, the *noumena* hypothesised by Kant. Derrida continues:

> Unlike the three other terms of the two correlations (*noesis-noema*, *morphe-hule*) this non-reality, this intentional but *non-real* inclusion of the noematic correlate is neither 'in' the world nor 'in' consciousness. But this is precisely the condition of any experience, any objectivity, any phenomenality, namely of any noetico-noematic correlation, whether originary or modified. It is no longer regional. Without the non-real inclusion of this intentional component (therefore inclusive and non-inclusive inclusion: the *noema* is included without being a part) one could not speak of any manifestation, of any phenomenality, in general (that being-for-a-consciousness, that appearing appearance which is neither consciousness nor the being that appears to it).
>
> (Ibid.)

Derrida then poses the questions:

> Is not such an 'irreality', *its independence both* in relation to the world *and* in relation to the *real* stuff of egological subjectivity, the very place of apparition, the essential, general, non-regional possibility of the specter? Is it not also what inscribes the possibility of the other and of mourning right onto the phenomenality of the phenomenon?
>
> (Ibid.)

Originary mourning emerges as the necessary unfoundedness of phenomena in any determinate determining *noumenon*, or being-in-itself. In this sense the phenomenon of Husserl's phenomenology is the prosthetic appearing out of which the thetic possibility arrives. This then is a creation of objectivity *ex nihilo* in the strong sense of there being no actual instance, in advance, but only after the fact of its appearing.

The dialogue *Sauf le nom* is a re-presentation of a dialogue, previously entitled *Post-scriptum*, and subtitled *Aporias*, ways and voices, and it is not fanciful to identify Aristotle, Heidegger and Husserl as implicitly invoked in these three terms; aporia, way, and voice. This dialogue was appended to the papers and discussion, in *Derrida and Negative Theology* (1992), arising from a conference in Calgary, Canada, which Derrida was, suitably enough, unable to attend. The collection begins with reprints in translation of two papers by Derrida, delivered in France, in 1980, and in Jerusalem in 1986. The first, 'Of an Apocalyptic Tone Newly Adopted in Philosophy', retrieves elements from the writings of Levinas, Heidegger and Kant, against the backdrop of a questioning of the effects of imposing the writings of New Testament Greek on the Hebrew, semitic traditions of eschatology. The second, 'How to Avoid Speaking: Denials' (1986) responds to, amongst other strands of thought, Jean-Luc Marion's recently published *God without Being* (1983), which dared to propose the theology without being, imagined by Martin Heidegger at a seminar in Zurich in 1951.[18] In the latter, Derrida adduces Heidegger's own distinction between theiology and theology to unsettle the notion of a theology without being, and also to multiply the notions of negative theology here in play, unsettling any attempt to transform negative theology all the same into a positive affirmation of doctrine. He here introduces a notion of a topolitology of the secret, an analysis of the place of the threshold, which permits the secret to be marked but left undivulged.

The dialogue *Sauf le nom* stages a double dialogue, one between Angelus Silesius and St Augustine, concerning what may be said of, and to God; and another between this dialogue, and moments of negative theology, and of political theology, at work in the thought of Edmund Husserl and Martin Heidegger. This brings out the relation between phenomenology and a certain inheritance of negative theology, as well as giving the name 'political theology' to the question of politics, for the phenomenologies which continue to affirm a link between a Greek beginning and the word 'philosophy'. It should perhaps be pointed out that those phenomenologies which do not raise the question of a Greek inheritance are those perhaps which even more clearly reaffirm that priority. This would be a version of the encrypting of a secret inheritance within traditions and practices, of which the bearers themselves are unaware: the condition of the Marrano. This 'transcendental contraband' is of ongoing concern to Derrida, from its naming in *Glas: What Remains of Absolute Knowledge* (1974) down into the postscript to *Donner la mort* (1999). It is a contraband which recalls Benjamin's conception of 'secret cargo'.[19]

Sauf le nom is published a second time under the title *On the name* (1993) between two further essays, 'Passions', first printed at the beginning of *Derrida: A Critical Reader*, and *Khora*, first printed in French in 1987.[20] 'Passions' considers the paradoxical expectation that a writer be able to pronounce authoritatively in relation to critiques of their own texts, to be able to speak, in the name of one's writings. It opens out an apophatic mode of language 'which is not necessarily dependent on negative theology, even if it makes it possible too' (p. 24). It considers the grammar of the phrase: 'There is something secret', arriving in the phrase: 'There, there is no longer time nor place.' *Khora* is invoked already in 'How to Avoid Speaking: Denials' as marking one of the three paradigms for discussion: the Greek of Plato's *Timaeus*; the Christian paradigm, of Eckhart and Dionysus; and Heidegger's reopening of the distinctions between onto-theology, theology, as meditation on God, and theiology, as accounts of the appearances of divinity. Derrida marks how *khora* is evoked by Plato as having a structure prior to that of the eternity of the intelligible paradigms, a temporality prior to eternity:

> Under the name of *khora*, the place belongs neither to the sensible nor to the intelligible, neither to becoming nor to non-being (the *khora* is never described as a void) nor to Being: according to Plato, the quantity or quality of Being are measured against its intelligibility. All the aporias which Plato makes no effort to hide, would signify that *there* is something that is neither a being nor a nothingness; something that no dialectic, participatory schema or analogy would allow one to rearticulate together with any philosopheme whatsoever.
>
> (D: 1989, HA p. 105)

For Derrida the term, *khora* above all poses the question of translation, as indeed does the title, *Sauf le nom*.

This title has at least a double edge. Bradley in his study *Negative Theology and Modern French Philosophy* (2004) reads it to show how Derrida, by retrieving the name of negative theology from any single determinate meaning, opens it out as the mark for a thinking of 'democracy to come'.[21] Derrida's dialogue questions the institution of a corpus of texts, gathered under the title 'negative theology'. One of the voices in the dialogue remarks:

> What we are identifying under these two words, today, isn't it first of all a corpus, at once open and closed, given, well-ordered, a set of statements recognizable either by their family resemblance or because they come under a regular logico-discursive type whose recurrence lends itself to a formalization.
>
> (D: 1993, ON p. 50)

When negative theology is understood neither as a body of texts nor as a set of prohibitions, it can be understood as a practice of religious observance, like prayer and ritual. Bradley's discussion of the encounter between Foucault and Derrida concerning madness, suggests that Foucault's technologies of the self provide a way of classifying one of the effects of negative theology as a practice, whereby the self is brought to surrender both the complacencies of human self-centredness and the naivety of an uncritically adopted religious practice. Negative theology and Christian mysticisms then emerge as sites at which the unauthorised and even in some cases women have asserted a right to a Christian experience, unmediated by the priestly and magisterial authorities of the day.

Bradley remarks how this line of enquiry is brought to the fore by Julia Kristeva and Grace Jantzens, in their attention to the mysticisms of some medieval women. It is important to recognise this strand of contestation within Christianity, while raising an even more pressing consideration: whether a distinction between theology and religion, reaffirmed in the distinctions between mysticism and liturgy, conversion and confession and in Kant's writings, necessarily Romanise and Latinise religion in the ways that Derrida's readings suggest. The second strand of discussion, which invokes phenomenology as a negative theology without the name, opens out a question to the political commitments of transcendental phenomenology, while simultaneously revealing that the political theology of negative theology, in affirming a democracy to come, may still implicitly privilege Christianity, and a Christian empire building, over other religions and other political programmes. Thus Derrida's question to Husserl and to his transcendental philosophy may also be posed to this conception of democracy. The epigraph at the head of this chapter suggests that transcendental phenomenology, as distinct from the phenomenology of Levinas, or Heidegger, or indeed Hegel, can be understood in its performance of the *epokhe* concerning existing entities, to refuse to engage in predication about what here is, instead exploring the reliability of meaning and of language for its expression.

The double dialogue in *Sauf le nom* can be read to show a certain conversion of negative theology into a political theology, tracing out a genealogy for components of a rethinking of democracy, into the thinking of a 'democracy to come', through a shared formation in strands of thinking associated with Christianity. Thus Derrida attaches his notion of the 'to-come' to this political-theological genealogy, detaching it both from Blanchot's conception of the Book, which is on its way, and from Levinas' dedication of an impossible thinking of time, to his God.[22] This intensifies the question of the relation of Derrida's thinking to the various traditions of Christianity, which he addresses through a reading of Jan Patocka in *The Gift of Death* (1992), and of his relation to the imposition of Christian readings on Jewish texts. *Aporias: Dying – Awaiting (One Another at) the 'Limits of Truth'* (1993) ends with an invocation of the condition of the Marrano, in Portugal, the sect of Jews who remained committed to Judaism, while outwardly conforming, and, in due course, losing contact with all but the name of their religion. This, he suggests, is the condition of the religious in secular society, and of Jews in the Christianised world of politics, and it adds force to his analyses of the secret at the foundation of community order, analysed in the opening sections of *The Gift of Death*, but not only there.

Towards the end of this doubled dialogue, conducted in *Sauf le nom*, between the two anonymised voices, Derrida writes:

> Take the example of democracy, of the idea of democracy, of democracy to come (neither the Idea in the Kantian sense, nor the current limited, and determined concept of democracy, but democracy as the inheritance of a

> promise). Its path passes perhaps today in the world through (across) the aporias of negative theology, that we just analysed so schematically.
>
> (D: 1993, ON p. 83)

The implications for a thinking of time, of a paradoxical inheritance of Christianity, in the notion of democracy, and of negative theology, in phenomenology, are to be traced out in this chapter, for which differences between aporia and paradox will be introduced as differences between a thinking with and a thinking without doctrinal commitment. The inheritance of this promise is what Derrida seeks to articulate by separating off the promise from Kant's analyses, and by detaching the messianic from messianism, detaching the thought of a fulfilled promise, from its doctrinal content as either Jewish or Christian, or as the event of an appropriation of the Jewish thought in Christianity, or indeed in Heidegger's notion of *Ereignis*.

That thinking is predicated on a certain logic of a necessity, which is necessarily lacking. This is introduced in *Sauf le nom*, in terms of a notion of a double abyss, with two distinct temporalities. Derrida's dialogue is speaking of 'A name of God':

> Each thing, each being, you, me, the other, each X, each name and each name of God can become the example of other substitutable X's. A process of absolute formalization. Any other is wholly other (*Tout autre est tout autre*) (trans. mod.). A name of God in a tongue, a phrase, a prayer, becomes an example of the name and of names of God, then of names in general. *It is necessary* (*il faut*) to choose *the best* of the examples . . .
>
> (D: 1993, ON p. 76)

The second voice then responds, and is in turn responded to:

> – *Il faut* does not only mean it is necessary but, in French, etymologically, 'it lacks' or 'is wanting'. The lack or lapse is never far away.
>
> – This exemplarism joins and disjoins at once, dislocates the best as the indifferent, the best as well as the indifferent: *on one side*, on one way, a profound and abyssal eternity, fundamental but accessible to messianism in general, to the teleo-eschatological narrative and to a certain experience of historical (or historial) revelation; *on the other side*, on the other way, the non-temporality of an abyss without bottom or surface, an absolute impassibility (neither life nor death) that gives rise to everything that it is not. In fact, two abysses.
>
> (D: 1993, ON pp. 76–77)

Thus Derrida indicates a layer of thinking, of an absolute impassibility, that of neither life nor death, beyond the teleo-eschatological narrative which he detects both in the historicity and historicality of Husserl and Heidegger, and indeed in the writings of Levinas and Kant.

The attempt to release a term, *khora*, from Plato's *Timaeus*, in order to articulate this further layer, as though reaching towards a determination beyond the limitations imposed by a certain European-ness, is a gesture which reveals the problem of an inescapability of the restricted and restricting nature of any inheritance. For this description of an absolute impassibility is thematised in relation to the retrieval of a certain term, within a quite specific tradition and transmission of enquiry, with its own conditions of translatability into the modern idioms in which it is both written and heard, when Derrida thus invokes it. It is for this reason that, while it might appear that in the late 1960s and early 1970s Derrida departed from the sphere of Husserl's phenomenology, nevertheless transcendental phenomenology is still in play in these later considerations of time, of inheritance, and of a certain genealogy, with the possibility of testifying both to that inheritance and to what is inherited, irrespective of its specific conditions. This attempt to go beyond its own restrictions Derrida understands as the distinctive move of philosophy, as performed by Kant and Husserl, and indeed in Aristotle's treatises and the dialogues of Plato:

> It is just this singular exemplarism that at once roots and uproots the idiom. Each idiom, (for example Greek onto-theology or Christian revelation) can testify for itself and for what it is not (not yet or forever) without this value of testimony (martyrdom) being itself totally determined by the inside of the idiom (Christian martyrdom for example). There, in this testimony offered not to oneself but to the other, is produced the horizon of translatability – then of friendship, of universal community, of European decentering, beyond the level of *philia*, of charity, of everything that can be associated with them, even beyond the European interpretation of the name of Europe.
>
> (D: 1993, ON p. 77)

The level of *philia*, it should be understood, is split between the restricted friendship of Greek ethics, and the supposedly unrestricted scope of Christianity and its political double, the brotherhood of man. The doubling of Christian revelation in the postponement in Judaism, and the doubling of a Greek origin for philosophy in the attempt to provide it with a universal, neutral jurisprudential conception of truth and justice, subject only to a language of the law court, reveals that any supposed absolute source is absolute only within the sphere of its own articulation: Christian discourse, Jewish law, Greek philosophy, Roman law. Benjamin's attempt to articulate a moment of divine violence, in advance of any affirmation or disruption of systems of law, in law making and law breaking, prompts in turn a gesture of violent interpretation in the name of the name, in 'The Force of Law: The "Mystical Foundation" of Authority' (1994). This moment of violence will be brought into conjunction with differences between Benjamin and Derrida concerning language and translation. This is the site of their most violent encounter, but does not remove that relation from the sphere of the brotherhood here invoked, which is marked from the beginning by the relation between Cain and Abel.

In *Sauf le nom*, Derrida introduces the name 'Husserl' into a discussion of how it is possible to recognise negative theology as negative theology, granted it is the discourse which is empty of determinacy. Derrida indicates that there is a danger of a certain mechanical organisation of texts, as belonging to such a corpus, accepting a tradition of interpretation, rather than determining a concept of their commonality in each occasion of invoking negative theology:

> All the more mechanizable and easily reproducible, falsifiable, exposed to forgery and counterfeit since the statement of negative theology empties itself by definition, by vocation, of all intuitive plenitude. *Kenosis* of discourse. If a phenomenological type of rule is followed for distinguishing between a full intuition and an empty or symbolic intending (*visée*) forgetful of the originary perception supporting it, then the apophatic statements are, must be on the side of the empty and then of mechanical indeed purely verbal, repetition of phrases without actual or full intentional meaning. Apophatic statements represent what Husserl identifies as the moment of *crisis* (forgetting of the full and originary intuition, empty functioning of symbolic language, objectivism etc.).
>
> (D: 1993, ON p. 50)

Suitably enough this remark deploys the New Testament Greek notion of *kenosis*, the mode of emptying out of language into an actualisation in the world, rendering indeterminate the distinction between word and world. For if there were truly a *kenosis*, there would be no discursive activity here to classify.

This remark also rehearses the distinction, drawn by Husserl, between authentic fulfilments of meanings and inauthentic repetitions. The remark continues:

> But in revealing the originary and final necessity of this crisis, in denouncing from the language of crisis the snares of intuitive consciousness and of phenomenology, they destabilise the very axiomatics of the phenomenological, which is also the ontological and transcendental, critique. Emptiness is essential and necessary to them. If they guard against this, it is through the moment of prayer or the hymn. But this protective moment remains structurally exterior to the purely apophatic instance, that is, to *negative* theology as such, if there is any in the strict sense, which can at times be doubted.
>
> (Ibid.)

The apophatic is that which either declines to affirm the positive attribute thereby implicitly assigning the negation of the attribute, or more cogently denies the very possibility of predication at all. This then sets out the apophatic and the kataphatic as converse movements, the one moving away from and the other moving towards the supposition that predication in the indicative tenses can retain some grammatical privilege:

> The value, the *evaluation* of the quality of intensity, or of the force of events of negative theology would then result from this *relation* that articulates *this* void (*vide*) in the plenitude of a prayer of an attribution (theo-logical, theio-logical, or onto-logical) negated (*niée*) let's say de-negated (*deniée*). The criterion is the measure of a *relation* (and this relation is stretched between two poles one of which must be that of positivity de-negated).
>
> (Ibid.)

This remark indicates a problem with the double work done by the English term 'deny', which both negates and suspends the negation. It also imposes the necessity to return to the discussion in 'How to Avoid Speaking: Refusing Negation', of Heidegger's distinction between theology and theiology, before returning to the notion of crisis and to the different responses of Derrida and Benjamin to a crisis of foundations for a thinking of emancipation, which brings into focus a difference concerning the thinking of time, in relation to religion and theology.

4.2 'There is the secret'

> Capitalism is probably the first case of a cult that produces guilt rather than atonement. In this respect, its religious system exists in the downfall of a monstrous movement. A monstrous consciousness of guilt, unable to find atonement, reaches for the cult not to find atonement, but rather to make the guilt universal – to hammer it into the conscious mind and finally and above all to include God Himself in this guilt, so as to finally interest Him in atonement.
>
> (Benjamin SW1, pp. 288–91)

This section of Walter Benjamin's early notes 'Capitalism as Religion' is quoted by Werner Hamacher, in his essay 'Guilt History: Benjamin's Sketch "Capitalism as Religion"' (2002) pp. 100–101.[1] The hypothesis is that religion is emptied out of its redemptive potential, in its figuration as capitalism. It is this redemptive potential which Benjamin rescues by inverting theology. This releases the concepts of redemptive time from their consolidation within religious forms and permits Benjamin to think the weak messianic force of an interruption of the various histories of tyranny and persecution. This of course presents the temptation to construe an antinomy between Derrida and Benjamin, as thinking contradictory theses, concerning theology without religion and religion without theology.[2] I shall show how the encounter between them is a great deal more violent than any such antinomial thinking would permit. This suggests that antinomy and paradox belong to a domain in which conceptual analysis and membership of faith communities are still compatible, whereas for both Derrida and Benjamin, the aporetics of European history and the corresponding aporetics of thinking an arrival of time in history, and the arrival of history in time, disjoin conceptual analysis and the condition of such membership, while leaving their traces the one on the other. The hypothesis that in capitalism, Christian theology realises itself must be left to one side.[3] In his remarkable analyses, Hamacher hypothesises that the *taxis*, or positing of time, in the Anaximander Fragment displaces and subordinates the time of human history, which is construed as merely a duration of awaiting the arrival of a last judgment:

> The time of history, ethical time is thus interpreted in Anaximander's sentence as a normative time of inculpation and expiation. Whatever enters

> this *taxis* of time is thereby already guilty and can only become excused by perishing.
>
> According to the thesis of Anaximander, time is the schema of guilt and retribution: The injustice committed by the progress of time occurs, however, like its remediation, unfreely. This time is therefore that of a guilt- and debt-continuum, continually advancing without a gap in its eternal recurrences. But it is not the time of history.
>
> (*Diacritics* 32.3–4, p. 82)

This analysis permits the formation of a hypothesis that Benjamin and Derrida, despite or perhaps because of the violence of their encounter, can be read as contributing to a formulation of a time of discontinuity, in which mortals are freed from these guilt- and debt continua. This suggests that underpinning natural scientific and everyday conceptions of time and underpinning the disputes between Levinas and Blanchot, Heidegger and Husserl on the inheritance of a phenomenological account of time, there is a conflict between two opposed conceptions of a relation between ethics and time.[4] There is this normative time of a Last Judgment, in which guilt, debt and capitalism are inextricably interlocked, as the destiny of a *mondialatinisation*. There is also, by contrast, an account of time as disrupted both by the *a-venir*, as that which arrives out of a future not thus foreclosed by the vision of a final reckoning, and by the weak messianic force, hypothesised by Benjamin, in which the past is brought back into the present.

An agreement between Benjamin and Derrida could then be discerned underneath the apparent falling out, in an analysis that this time of destiny and of a natural eternal recurrence, must be interrupted, if there is to be an experience of freedom. The difference would be the manner in which, in order to think this interruption, Benjamin mobilises theological concepts violently detached from their religious origins, whereas Derrida makes use of a form of Husserl's techniques of bracketing and suspending content, to form the notion of the messianic without messianism. The line of development between Benjamin's early essay 'Critique of Violence' (1921), which Derrida reads in 'The Force of Law: The "Mystical Foundation of Authority"' (1989), and the late series of aphorisms 'On the Concept of History' (1940) reveals both a continuity and a sharpening of the impact of detaching these concepts from their liturgical and ritual deployments. Hamacher, in another essay, '"Now": Walter Benjamin on Historical Time', reads the latter to mark up a notion of time as that of the missed opportunity. The time of translation, by contrast, is that of the retrieval of that opportunity. Hamacher remarks: 'The Messiah who is supposed to rescue the missed possibilities of history may himself be missed.'[5] This marks a time in which history may fail to take place, and there occurs instead the disaster of continuation, which for Benjamin is the form of the idea of progress. This is the mode of 'living on' deployed now in relation not to the writings of dead authors, but to the processes of history, similarly understood as cut loose from any analysis of them as resulting from intended actions.

As Hamacher analyses it, there are the two models of time, as subordinated respectively to a first cause, and to chance; and there is this third option of thinking time as independent of either of these, as emergent from local sedimentations of flows and movements, which for a while give rise to an appearance of an unlimited *taxis* of time, but which subside back into an overflowing of any such positing and tempering. Well tempered time maintains a constant rhythm; distempered time is subject to interruptions which bring flow to a standstill, with the potential for redirection. There are modes of thinking the relation between these two times, with Derrida, as out of joint, and unhingeing, or with Jean-Luc Nancy as syncopation, or, between the two, as the flutter of the heartbeat which has its own necessary limitations. Genesis, as becoming, is no longer to be understood as 'a guilt (*adikia*) that must be expiated in perishing' (*Diacritics* 32.3–4, p. 81). Becoming is to be released from any such pairing with perishing, and is thus detached from the circumscription of organicism, where what comes into being is subject to a process of ripening and decay. Genesis can be rethought as passive synthesis, as *a-venir* and prosthesis, as writing and the *circonfession* which rotates around the thought of the other, from which it takes inspiration, and away from which it necessarily strays. *Circonfession* thus traces the movement of an aberrant ellipse, which does not return to its previous positions. The privilege to the regular evenness of rhythm ascribed to the flow of time is thus brought into question and the slowing down and speeding up, the hesitation and precipitation of Derrida's mode of writing become part of a diagnosis of the structure of time which gets hidden when the emphasis is put on regularity, and regularisation, on counting and reckoning, rather than on decisions and destinies. This structure of time is excavated by Derrida in relation to the mystery and concealment at the heart of religious ritual and in their transmission. In the notion of a *topolitology of the secret*, introduced in 'How to Avoid Speaking: Denials' (1986), and implicitly in *The Gift of Death* (1992, 1999), Derrida analyses place, in relation to time, and in relation to concealments which are a feature both of the structure of consciousness, which cannot be aware simultaneously of all its contents, and of the structure of the transmission of mystery cults. Topolitology thus is the study of the place of the threshold or border, which has the form of the secret, and which is formed as a result of the manner in which secrets are transmitted. This generates the possibility of rethinking time as a many-dimensional temporality, inflected through its various points of origin, forms of articulation and modes of attestation, as irreducibly multiple, and discontinuous, in line with the notions of originary technicity and originary mourning.

This topolitology can be extended to provide an account of the border or place at which the encounter between Benjamin and Derrida does and does not take place. My hypothesis in reading these various texts and in setting up an encounter between Benjamin and Derrida is that religious discussion, texts and sectarianism keep open questions of identity and violence which are closed off in the presumption that the only political system worthy of the name is democracy and that there is no problem about what democracy might consist in. Thus religion turns into the crypt within which the undiscussable aspects of politics are transmitted across the regimes of repressive toleration, as their 'transcendental contraband'. There are thus at least four layers in the deployment of a notion of topolitology. There is its deployment by Derrida to describe the effects on time and temporality of the manner in which secrets are transmitted between people and across generations. There is its role as an articulation of the structure of consciousness in which layers may, as in Husserl's analyses of passive synthesis, necessarily remain occluded. There is this role, as a description of the encounter and non-encounter between Derrida and Benjamin, on the questions, how to think history and time, violence and translation. There is also a role in revealing connections between a supposedly secular domain of democratic and political aspiration, with its Greek roots, and the religious theological domains of faith and ritual. The secret of democracy is its role in containing the violence of religious sectarianism, while the secret of religion is the unrealisable status of the aspiration, in democracy, to universal community, with universalisable prescriptions. The promise of a formal right to participation in democratic process, secured supposedly by a system of voting, with their various inadequacies and artifices, never provides the determinacy of identity and community offered by the sect.

Derrida develops the figure of an auto-immunisation, whereby an organism destroys itself by destroying what supposedly threatens it.[6] He writes in the earlier essay 'Faith and Knowledge':

> This dignity of life can only subsist beyond the present living being. Whence, transcendence, fetishism, and spectrality; whence the religiosity of religion. This excess above and beyond the living, whose life only has absolute value by being worth more than life, more than itself – this in short is what opens the space of death that is linked to the automaton (exemplarily 'phallic'), to technics, the machine, the prosthesis: in a word, to the dimensions of the auto-immune, and self-sacrificial supplementarity, to this death-drive that is silently at work in every community, every *auto-co-immunity*, constituting it as such in its iterability, its heritage, its spectral tradition.
>
> (D: 2002, AR p. 87)[7]

This he calls the principle of sacrificial self-destruction, ruining the principle of self-protection. This is identified by Derrida in the second essay in *Rogues*, as the key to rethinking an auto-destruction of reason, as an inward looking practice of democracy, which identifies its enemy as necessarily anti-democratic. Analysis of these processes of destructive auto-immunisation makes it possible to re-open the process of defining the separation of powers, supposed necessary for securing the future of democracies, in a move beyond the distinctions usually drawn between legislative, executive and administrative instances. For this may not exhaust the various distinct aspects in play, serving rather to conceal the distinction between ecclesial and spiritual powers, and their claim to take precedence over the secular or 'temporal'.

The notion of the border analysed in 'Faith and Knowledge' follows the lines of the threefold distinctions between borders, already marked in *Aporias: Dying – Awaiting (One Another at) the 'Limits of Truth'* as literal, concerning national states, practical, concerning engagement in distinct activities of law, religion, economy etc., and conceptual:

> *(24) The surge of 'Islam' will neither be understood nor answered as long as the exterior and interior of this borderline place have not been called into question: as long as one settles for an internal explanation (interior to the history of faith, of religion, of languages or cultures as such) as long as one does not define the passageway between this interior and all the apparently exterior dimensions (techno-scientific, tele-biotechnological, which is to say also political, and socioeconomic, etc.).*
>
> (D: 2002, AR p. 58)

Derrida's title on this occasion, as in 1980, when he wrote 'Of an Apocalyptic Tone' mimes the title of text by Kant, in this case *Religion within the Bounds of Reason Alone* (1794),[8] in which Kant analyses the conceptual determinations through which to grasp the efficacy of religion. Derrida identifies how Kant introduces a series of additions, called *Parerga*, a term to which he has already drawn attention in his reading of Kant's *Critique of Judgment* (1791) in *Truth in Painting* (1978). These *parerga* provide boundaries which are also disruptive addenda to the matter already discussed:

> *This definition of reflecting* faith *appears in the first of the four* Parerga *added at the end of each section of* Religion within the Limits of Reason Alone. *These* Parerga *are not integral parts of the book: they 'do not belong within' 'religion in the limits of pure reason', they 'border upon' it. I stress this for reasons that are in part theo-topological, even theo-architectonic: these* Parerga *situate perhaps the fringe where we might be able, today, to inscribe our reflections. All the more since the first* Parergon *added in the second edition, thereby defines the secondary task* (parergon) *which concerning what is morally indisputable would consist in surmounting all the difficulties connected to transcendent questions.*
>
> (D: 2002, AR p. 52)

The *parergon*, plural *parerga*, are additions to the work, the *ergon*, which reveal something which has not yet been said or performed by the work. In *Truth in Painting*, Derrida suggests that Kant's attempt to foreclose certain features of artworks as subordinate and subsidiary considerations, under this title *parergon*, must be unsuccessful.[9] Derrida then remarks:

> *We will have to limit ourselves to an indication of the title of this programme and first of the criteria (nature/supernatural, internal/external, theoretical elucidation/practical action, constative/performative): (a) the allegedly internal experience (of the effects of grace): the fanaticism or enthusiasm of the illuminated; (b) the allegedly external experience of the miraculous: superstition (*Aberglaube*), (c) alleged elucidation of the*

> *understanding in the consideration of the supernatural (secrets,* Geheimnisse*): illuminatism, the frenzy of initiates; (d) the risky attempt of acting upon the supernatural (means of obtaining grace): thaumaturgy.*
>
> (Ibid.)

In each pairing the first is according to Kant the acceptable practice and the second its corruption and distortion.

Derrida then makes the link to Marx's diagnosis of religion as ideology, on the basis of his analyses of capitalism and revolution :

> *When Marx holds the critique of religion to be the premise of all ideology-critique, when he holds religion to be ideology par excellence, even for the matrix of all ideology and of the very movement of fetishization, does his position not fall, whether he would have wanted it or not, within the parergonal framework of this kind of rational criticism? Or rather, more plausible but also more difficult to demonstrate, does he not already deconstruct the fundamentally Christian axiomatics of Kant?*
>
> (D: 2002, AR p. 52)

The difficulty may, however, rather be that on the contrary the Christian axiomatics of Kant's critique can be shown to hold the Marxian critique in place. Truly startling in this text is the patient analysis of Kant's moral philosophy, in the context of the distinction Kant draws between religions of cult and religions of morality. This reveals Kant to suppose that Christianity, and indeed Protestantism, is the only truly moral religion. Derrida writes:

> *The unconditional universality of the categorical imperative is evangelical. The moral law inscribes itself at the bottom of our hearts like a memory of the Passion. When it addresses us, it either speaks the idiom of the Christian – or it is silent.*
>
> (D: 2002, AR p. 50)

Derrida notes the conflict between Christianity and the other two monotheisms, with more than a trace of the thought that only Christianity still maintains a living relation to a previous paganism, and to pagan cults of sacrifice.

Derrida reads Kant's text as providing a grounding for morality, even in the condition of abandonment by God.

> *Kant recurs to the logic of a simple* principle *that we visited a moment ago verbatim: in order to conduct oneself in a moral manner, one must act as though God did not exist or no longer concerned himself with our salvation. This shows who is moral and who is therefore Christian, assuming that a Christian owes it to himself to be moral: no longer turn towards God at the moment of action in good faith: act as though God had abandoned us.*
>
> (D: 2002, AR pp. 50–51)

Derrida hypothesises the conclusion: '*That Christianity is the death of God thus announced and recalled by Kant to the modernity of the Enlightenment?*' Thus the separation between a death of the God of the ontologists, achieved by Kant, is matched here by a death of God in the moral sphere, neither of which are required any more to secure the well-orderedness of enquiry in natural and moral science. Derrida goes on to surmise that there is a resistance in Judaism and in Islam to the attempt thus to remove essential reference to God from cosmology and moral understanding:

> *Judaism and Islam would thus be perhaps the last two monotheisms to revolt against everything that, in the Christianizing of our world, signifies the death of God, death in God, two non-pagan monotheisms, that do not accept death any more than multiplicity (the Passion, the Trinity etc.) two monotheisms still alien enough at the heart of Graeco-Christian, Pagano-Christian Europe, alienating themselves from a Europe that signifies the death of God, by recalling at all costs that 'monotheism' signifies no less faith in the One, and in the living One, than belief in a single God.*
>
> (D: 2002, AR p. 51)

Thus to require the adherents of Islam and Judaism to displace the role of faith from the formation of states, founded in their name, to require an elimination of the theocratic element is to challenge theocracy in the name of a theocracy which pretends not to be one. The appearance of secularism in secular states inherits a specific, Christian mode of separation, which on closer inspection turns out to be not so very separate, granted the banning of the display of Christian symbols in French schools; the constitutional role of the Church of England; and the chances of a non-churchgoer being elected president of the United States.

Derrida marks up a strand of response to this, running from Nietzsche to Heidegger, in which the separation of morality from Christian commitments is signalled as a necessity, if philosophy is to arrive at a non-sectarian foundation, independent of religious commitment:

> *This would in principle allow for the repetition of the Nietzschean genealogy of morals, but dechristianizing it where necessary and extirpating whatever Christian vestiges it still might contain. A strategy all the more involuted and necessary for a Heidegger who seems unable to stop either settling accounts with Christianity, or distancing himself from it – with all the more violence in so far as it is already too late, perhaps, for him to deny certain proto-Christian motifs in the ontological repetition and existential analytics.*
>
> (D: 2002, AR p. 51)

The *mondialatinisation* which he seeks to analyse is thus '(*this strange alliance of Christianity as the experience of the death of God and tele-technoscientific capitalism*)' (pp. 51–52). The writings of Marx and Nietzsche thus can play some role in assisting the analysis but cannot be presumed to be entirely independent from that which is to be analysed, in the same way as Heidegger's enterprise of setting out a philosophical enquiry neutral with respect to a Christian, theological inheritance turns out to be more difficult than at first envisaged.

The response here to the writings of Derrida and of Benjamin thus seeks to juxtapose them against this backdrop, of a questioning of the supposed absence of theocracy in democratic states. The three topics for discussion are the notion of topolitology, theiology and the transition from translation to violence. Topolitology is the analysis of these borders, limits and crossings, of their transgression, disruption and re-affirmation. Theiology, marked up by both Derrida and Heidegger as distinct from theology, disrupts the transition from theology to either onto-theology or political theology. The differences between Derrida and Benjamin on meaning, language and translation reveal a connection between theorising translation and thinking the moment of violence, as distinct ways of thinking a disruption of time. For this disruption of a smooth ordering of time as thought by Derrida and Benjamin invokes distinctions between the divine and the mortal, and between religion and theology. Their responses to these questions are, however, the more open, and the more charged, granted their relations of self-distancing from two of the three monotheisms in question: Judaism and Christianity. Analysis of these differences opens out on to an understanding of a paradoxical temporality of discontinuities and uneven rhythmic structures, through which to reveal the uneasy alliance between a secularising politico-philosophical intent and this Christian inheritance. The task would be to show the Christian components in certain construals of political theology and to show the particularism underlying claims to universalism.

In 'How to Avoid Speaking: Denials', Derrida offers an incomplete reading of *The Mystic Theology* of Dionysius, which feeds into the later papers '*Sauf le nom*' and 'Faith and Knowledge'. It has already been remarked that this address is in part a response to the recent publication of Jean-Luc Marion's study, *God without Being* (1983).[10] It is also a response to the paper 'Of Divine Places', delivered by Jean-Luc Nancy in 1984, similarly responding to the writings of Jean-Luc Marion.[11] Derrida marks up a difference between God, and the place in which that God is supposed to reside, leading to the thought of a God who resides in no place: 'To gain access to this place is not yet to contemplate God. Even Moses must retreat. He receives this order from a place that is not place even if one of the names of God can sometimes designate place itself' (D: 1989, HA p. 91). The political theology derived from a thinking of a God who resides in one place, is of course distinct from a political theology derived from a thinking of a God which resides in no place. The former is of course still pagan. Such differences are the concern of a topolitology, expanded beyond its initial scope as a topolitology of the secret, which Derrida announces in this paper by making three moves. The first move points up the role of the order, or injunction, supposed to emanate from that God: '1: To separate oneself, to step aside, to withdraw with an elite, from the start this topolitology of the secret obeys an order' (D: 1989, HA p. 92). The second move suggests the complicity with certain practices in literary study, the attention to figuration and to rhetorical analysis, but indicates an additional charge once the significance of the use of language in describing God, as a form of distorting representation, is grasped:

> 2: In this topolitology of the secret, the figures or *places* of rhetoric are also political stratagems. The 'sacred symbols' the compositions, the signs and figures of the sacred discourse, the 'enigmas' and the 'typical symbols' are invented as 'shields' against the many. All of the anthropomorphic emotions which one attributes to God, the sorrows, the angers, the repentances, the curses, all negative moments – and even the 'sophistries' which He uses in the Scripture 'to evade His promises' – are nothing but 'sacred allegories' which one has had the audacity to use to represent God, projecting outward and multiplying the visible appearances of the mystery, dividing the unique and indivisible, figuring in multiple forms which has neither form nor figure.
>
> (Ibid. p. 93)

This remarks the deployments of rhetoric and allegory in sacred texts to present the unrepresentable marks the concept of the sublime, through which a transition from considering sacred to responding to profane writing takes place.[12] The link to a thinking of literature is made thus: 'It would suffice to doubt this promise or transgress this injunction in order to see an opening – and also a closing up itself – of the field of rhetoricity or even of literariness, the lawless law of fiction' (Ibid.).

The analysis of this divine promise continues:

> Since the promise is also an order, the allegorical veil becomes a political shield, the solid barrier of a social division; or if you prefer a *shibboleth.* One invents it to protect against access to a knowledge which remains in *itself* inaccessible, untransmissible, unteachable.
>
> (Ibid.)[13]

Derrida then remarks 'another political and pedagogical consequence: another institutional trait: the theologian must practice not a double language but a double inscription of his knowledge. Here Dionysus evokes a double tradition, a double mode of transmission' (D: 1989, HA p. 94). This is: 'on the one hand, unspeakable, secret, prohibited, reserved, inaccessible', and 'on the other hand, philosophic, demonstrative capable of being shown'. Concerning this tension, Derrida remarks: 'The critical question evidently becomes: How do these two modes relate to each other? What is the law of their reciprocal translation or of their hierarchy? What would be its institutional or political figure?' The paradox is underlined thus:

> At the crossing point of these two languages, each of which bears the silence of the other, a secret must and must not allow itself to be divulged. It can and it cannot do this. One must not divulge, but it is also necessary to make

> known or rather allow to be known this 'it is necessary', 'one must not' or 'it is necessary not to.'
>
> (D: 1989, HA p. 94)

This modality of the making-known of the necessary not making-known reveals a complex interiority of the 'perhaps' which emerges once it is inflected through the modality of the promise.

The third remark traces the duality in this notion of a lack of place, as both *atopics*, a lack of figuration, and *atopos*, that which is senseless. There are then linkages from the negativities of negative theology to the ban on prophecy in Judaism and on figuration in Islam. To mark this third folding of topolitology, of a questioning of topology, into an analysis of the formation of meaning, as predicated on silence, Derrida closes the first section of his paper, and opens a second section, with a typographical move, which of course is visible and effective only when read. It reads:

> II.
> We are still on the threshold.
>
> (Ibid. p. 96)

A thinking of this threshold is focussed by him in the following way:

> In the three stages that now await us, I have thought it necessary to privilege the experience of place. But already the word *experience* appears risky. The relation to the place about which I should speak will perhaps no longer have the form of experience – if this still assumes the encounter with or crossing over a presence.
>
> (Ibid.)

Experience will thus be transposed from its first inscription as an 'encounter with or crossing over of a presence' into an encounter with or failure to cross over to the place, and into a challenge, of the thought and self-constitution of the other, in this case invoked under the name 'Walter Benjamin'.

The three places to which Derrida proposes to appeal are those of the articulation of Plato's *khora*; the *via negativa* 'in its Christian stage'; and third,

> If I were not afraid of trying your patience, I would recall that which, in Heidegger's thinking, could resemble the most questioning legacy, both the most audacious and the most liberated repetition of the traditions I have just evoked. Here I will have to limit myself to a few landmarks.
>
> (Ibid. p. 122)

This section is introduced with the following *caveat*:

> I thus decided *not to speak* of negativity or of apophatic movements in, for example, the Jewish or Islamic traditions. To leave this immense place

> empty, and above all that which can connect such a name of God with the name of the Place to remain thus on the threshold – was this not the most consistent possible apophasis? Concerning that about which one cannot speak, isn't it best to remain silent? I let you answer this question. It is always entrusted to the other.
>
> (Ibid.)

Derrida introduces one set of distinctions, only to complicate it immediately:

> It is necessary to distinguish between, on the one hand, onto-theology or theiology, and on the other hand, theology. The former concerns the supreme being, the being *par excellence*, ultimate foundation or *causa sui* in its divinity. The second is a science of faith or of divine speech, such as it manifest s itself in revelation (*Offenbarung*). Heidegger again seems to distinguish between manifestation, the possibility of Being to reveal itself (*Offenbarkeit*), and on the other hand the revelation (*Offenbarung*) of the God of theology.
>
> (D: 1989, HA pp. 123–24)

The link from these distinctions to Derrida's own distinction between the messianic and messianism remains to be made.

From this encounter with Heidegger, then, I have selected out this distinction between theology and theiology, neglecting further discussion of what Derrida does and does not affirm here. To do justice to Derrida's questioning of Heidegger on the legacy of negative theology, and the refusal of negation, would however require reading this text, 'How to Avoid Speaking', back to back with its other half, *Of Spirit: Heidegger and the Question* (1987), in which Derrida traces out the vacillation in Heidegger's avoidance of the term *Geist*, in an analysis of the opening and closing of single and double quotation marks.[14] For this vacillation with respect to *Geist* is also a vacillation with respect to the Christian inheritance. In that second text Derrida explicitly draws attention to this need for a parallel reading, which cannot be done here. One further puzzle about the discussion in 'How to Avoid Speaking' does need remarking. Derrida draws attention to the *topos*, figure and fact, of the place, and of its suspension, in the thinking of negative theology and of *khora*. He remarks that, in his discussion of its three way-stations:

> Each time, problems are inevitable: *on the one hand*, the immense problem of figurative spatialization [both *in* speech or writing in the current sense and in the space *between* the current sense and the other of which the current sense is only the figure], and, *on the other hand* that of meaning and reference, and *finally*, that of the event in so far as it takes place.
>
> (Ibid. p. 97)

And each time he marks up how his response to the Greek tradition from which the notion of *khora* emerges, his response to the tradition of negative theology, in Christianity, and his reading of Heidegger must be cut short, or cut to fit the time available, as he addresses his listeners. The time of negative theology must fall into the time of the lecture; but it must also fall into the time of a certain chronology which understands as obvious a move back to and forwards from these Greek origins, to a Christian transmission, in New Testament Greek, in the Latin of Augustine, church Latin, the languages of the Low Country mystics, and its transposition through Luther's German Bible, to Heidegger. Theology can present itself only through the mediation of the languages which present themselves, and the appearing of the divine, theiology, reveals how even onto-theology is foundational always only if the mediation of the language in which it is articulated is forgotten, and only if theiology as an account of the appearances of the divine is restricted to the manifestations of an internally consistent monotheism.

The three readings of Benjamin offered by Derrida provide very different angles on any supposed encounter between them.[15] These encounters between Derrida and Benjamin also disrupt Derrida's reading of Heidegger, as the encounter with Benjamin is disrupted by his readings of Heidegger. For in 'Force of Law', Derrida surmises that Benjamin, Benjamin's protests to the contrary, is still too close to Heidegger's thought of an originary violence which marks human beings.

> *It is at that point in this text, in all its polysemic mobility and all its sources for reversal, seems to me finally to resemble too closely, to the point of specular fascination and vertigo the very thing against which one must act and think, do and speak. This text like many other by Benjamin is still too Heideggerian, too messianico-Marxist or archeo-eschatological for me.*
> (D: 2002, AR p. 298)

This misses the careful separation of the 'weak messianic force' of Benjamin's 'On the Concept of History' (1940), from the inscription of such eschatology in a teleology of fulfilled history. There is also a difference to be marked between Benjamin's notion of a bloodless violence of the arrival of a divine moment, and the violence invoked by Heidegger, in the challenge presented by the human to the non-human power of sending, in which is revealed the full power of an indifferent irruption of being into history. This Heidegger analyses at length in *Introduction to Metaphysics*, the lectures from 1935–36, when his deviation from his adopted Nazism begins to come to the fore. Derrida's presumption of an affinity between Heidegger and Benjamin here erases the marked differences between Heidegger's archaising references back to the plays of Sophocles, by contrasts to Benjamin's efforts to identify what is new and different about the work of art and its relation to human history, in for example, the contemporaneous essay 'The Work of Art in the Age of Its Mechanical Reproducibility' (1936).

The analysis of the human as the strangest of all is developed by Heidegger in this lecture series, *Introduction to Metaphysics*, through a reading of Sophocles' play *Antigone*.[16] Heidegger writes:

> As the breach for the revealing of Being which has been set to work in beings – the Dasein of historical humanity is an incident (*Zwischenfall*) the incident in which the violent powers of the released excessive violence of Being suddenly emerge and go to work as history.
>
> (*Introduction to Metaphysics* p. 174)

Heidegger's figure for such violence is, of course, Oedipus, whose hubris challenges the divine order. The context in which he delivered these lectures, however, gives the analysis a distinct edge, since it is hard not to hear the organisation of Nazi genocide at work in it. Benjamin's notion of destruction, marked up in the title of his essay from 1931, 'The Destructive Character', poses a challenge both to this and indeed to Derrida.[17] For Derrida's notion of deconstruction is more easily reconciled with the notion of destruction deployed by Heidegger in *Being and Time*, concerning a destruction of the history of ontology, than with the image conjured up by Benjamin's words:

> The destructive character sees nothing permanent. But for this very reason he sees ways everywhere. Where others encounter walls or mountains, there too he sees a way. But because he sees a way everywhere, he has to clear things from it everywhere. Because he sees ways everywhere, he always stands at a crossroads. No moment can know what the next will bring. What exists he reduces to rubble, not for the sake of rubble, but for that of the way leading through it.
>
> (Benjamin SW2, p. 542)

This notion of destruction might be thought to be that of *Being and Time* which proposes to take apart a history of ontology. Subsequently, however, Heidegger supposes that this history sets itself back in place, in the resistance of the language of metaphysics to such treatment, as a result of which *Being and Time* cannot arrive at its proposed result of rethinking being as time. Oedipus cannot inaugurate a new beginning in the twenty-first century, and for Benjamin, Heidegger's *Being and Time* is not the rethinking of time and history, needed to make sense of the twentieth century.

Benjamin marks a resolve to meet the Heidegger of *Being and Time* on his path to a theory of dialectical images, and 'to reduce him to rubble'. He writes to Scholem on 'April 25, 1930: We have come up with the plan in quite a small critical reading group this summer under the leadership of myself and Brecht to reduce Heidegger to rubble' (*Benjamin Briefe Bd 11*, p. 514).[18] This then is antipathy, not antinomy, and Benjamin proposes texts for analysis less self-defeating and less caught up in an archaisation of cultural struggle than those even of Hölderlin. He certainly does not propose a return to an engagement with Greek thinking as though no significant structural historical change has taken place since then. Instead of translating Sophocles into modern idiom, with Hölderlin and Heidegger, Benjamin reveals the resistance to translation of the idiom of Proust and of Baudelaire.[19] While there are profound differences between Derrida and Benjamin on the logic of translation, the chiasm of the religion without religion and of religion without theology, turns out on closer inspection to conceal a joint exploration of the possibility of a redemptive movement, without sacrifice, through which to break out of history. They share an analysis of history as nothing more than the monumentalised memories of those who live on. In the next section of this study I shall consider the resulting splitting apart of an analysis of time from an analysis of history.

The encounter between Benjamin and Derrida thus obeys two laws to which these enquiries have up until now only implicitly conformed: that such an encounter takes place in a number of occurrences, with a series of pre- and after-shocks, spread over several readings of several distinct texts; and that there are no encounters exclusive to two parties, but always third, fourth and fifth parties irreducibly implicated in whatever might be thought to take place. There is first the interruption of an encounter between Derrida and Benjamin by this third figure, Heidegger, who might be thought to be Benjamin's true enemy; there is second a disruption of the encounter between them, by Derrida's invocation of Adorno, who might be thought to be Benjamin's false friend; and third there is the disruption of any encounter between them by Derrida's reluctance to take on the full force of Levinas' affirmation of divine violence, to which Derrida is nevertheless well attuned, demonstratively from the writing of 'Violence and Metaphysics' on. The role of this fifth figure, Levinas, is perhaps the most significant, for there is a stronger clash between Levinas on meaning, and the possibility of translation, and Benjamin's account, than between any of the other possible pairings. Derrida's encounter with Benjamin in 2001 is stalled by the arrival of the name 'Adorno'; the encounter in 1989 is disrupted by the arrival of that of Heidegger. In 1984 it is the name 'Levinas' which silently intervenes, and with which I shall begin.

Des Tours de Babel, detours around, or concerning the Towers of Babel, derives from a lecture delivered in 1984, in which Derrida invokes the proper name 'Babel', as a place-holder for the untranslatable, proper text.[20] He indicates how Voltaire in his dictionary moves from remarking that 'Babel' means 'God the Father', to remarking how 'Babel' comes to mean confusion. Derrida then adduces:

> Babel: today we take it as a proper name. Indeed, but the proper name of what and of whom? At times that of a narrative text recounting a story (mythical, symbolic, allegorical; it matters little for the moment) as a story in which the proper name, which is then no longer the title of the narrative, names a tower or a city, but a tower or city that receives its name from an event during which YHWH 'proclaims his name'.
>
> (D: 2002, AR pp. 108–9)

This is indeed a conflict of proper names, whereby both town, and God, receive untranslatable names. From time to time, and certainly in the running commentary to 'Living On – Border Lines', it seems as though irreducible multiplicity is precisely what the most exacting learning of languages must demonstrate: that the slippage of one language simply cannot be translated into the slippage of another.[21] The movement of translation overlays a certain movement of transference of libidinal energy which again cannot be stabilised and made a determinate calculable quantity of energy or quality of meaning intending.

In the essay on translation, Derrida remarks the link between proper names and a certain untranslatability:

> Now, this proper name which already names at least three times and three different things also has, this is the whole point, as proper name the function of a common noun. This story recounts, among other things, the origin of the confusion of tongues, the irreducible multiplicity of idioms, the necessary and impossible task of translation, its necessity *as* impossibility.
>
> (D: 2002, AR p. 109)

This very Kantian formulation of a necessity is then to be contra-posed to the necessity explored by Benjamin of turning language into a system of necessary citation, whereby what is said names itself as its own thought content. Benjamin thus seeks to secure thought contents from the abrasions of articulation in natural languages.[22] For Derrida remarks that proper names do not belong to language, in the same sense as does the translatable:

> Now, in general one pays little attention to this fact: it is in translation that we most often read this narrative. And in this translation, the proper name retains a singular destiny, since it is not translated in its appearance as proper name. Now, a proper name as such remains forever untranslatable, a fact that may lead one to conclude that it does not strictly belong, for the same reason as the other words, to the language, to the system of the language, be it translated or translating.
>
> (Ibid.)

And he points up the manner in which the task, *Aufgabe*, of the translation is linked not to a transformation of the Kantian task of a critique of reason, but to a common sense notion of surrender, of giving up on an undertaking.

This, of course, is not at all Benjamin's understanding of his own title 'The Task of the Translator' (1923), as Derrida indicates by pursuing the thought that for Benjamin through translation there is achieved a living on (*Fortleben*) of the text. On this occasion, Derrida affirms that his engagement with Benjamin will be oblique: that what is called for is a reading of 'On Language as such and the Languages of Humanity' (1916) but that he has chosen to substitute a reading of the essay 'The Task of the Translator', although in fact he does then also double back to the earlier text.[23] This latter text, which preceded Benjamin's translation of Baudelaire's *Tableaux parisiens*, has the advantage, for Derrida, of posing the problem of a three-way translation, where there are words in the text from the language into which it is to be translated. This also takes place when Derrida's texts are translated from French into English or, more rarely, from English into French. This displaces the phrase, for it can no longer have the effect that it has had when quoted as a foreign, say, English, phrase in the French. This encounter with Benjamin is, however, rendered structurally incomplete by its role in mediating the relation between Derrida and the memory of Paul de Man, in a questioning of biography and mourning, of autobiography, and the writing of *mémoires* as the impossible attempt to meet up with oneself and another in a future which it is impossible to access: a future after one's death.[24] This raises the question of inheriting the task of critique, to which de Man makes an indispensable contribution, taking up Kant's critical philosophy through an affirmation of literary reading.[25] This returns for discussion in Part V.

The much commented text 'Force of Law' consists in a lecture as delivered in October 1989 to the Cardozo Law School, and a section distributed to the participants, and delivered in April 1990: 'On Benjamin's First Name'. In the latter there is to be found the unlikely complicity remarked on above between Benjamin and Heidegger. In the first half, Derrida puts forward the twin notions 'Deconstruction is justice' and 'Justice is undeconstructible':

> In the structure I am here describing law is essentially deconstructible, whether because it is founded, that is to say constructed, upon interpretable and transformable textual strata (and that is the history of law, its possible and necessary transformation, sometimes its amelioration) or because its ultimate foundation is by definition unfounded. The fact that law is deconstructible is not bad news. One may even find in this the political chance of all historical progress. But the paradox that I would like to submit for discussion is the following: it is this deconstructible structure of law or, if you prefer, of justice as law, that also ensures the possibility of deconstruction. Justice, in itself, if such a thing exist, outside or beyond the law, is not deconstructible. No more than deconstruction itself, if such a thing exist. *Deconstruction is justice.*
>
> (D: 2002, AR pp. 242–43)

This mode of analysis, however, is vulnerable to a critique from Benjamin's attempts to reveal an anterior splitting not just between law, legality and a concept of justice, but within a concept of justice itself, between an irrecuperable divine moment, its human markings, and a notion of natural history. This reinstallation at the origin of meaning between a divine and a human instance Derrida will seek to resist, for the reason which becomes clearer in the work on animality: that it tends to lay emphasis on a connection from the human to divinity, rather than accepting the framing, preferred by Derrida, between animality, as organism, and the human partiality for machines.

Natural history reveals humanity as animal, and it is to this that Derrida turns in the short version of 'The Animal Which Therefore I Am'. A splitting of justice between a naturalised notion, in which all that comes into being must pass away again, and a specifically Benjaminian notion of justice, which detaches meaning from history, is brought into focus by the reading Hamacher gives of Benjamin's sketch 'Capitalism as Religion'. This installation of a site for affirming meaningfulness independent of a passage of time as history is contrary to the movements marked up in the thinking of *différance*, which Derrida redeploys in the context of these analyses of law and justice, and which remains the main structure for his thinking time, as a moment suspending both time as chronology and time as historical narrative, but rejecting neither. Derrida invokes *différance* to mark differences between force, violence, and legitimate authority which are both invoked and covered over in the various terms translated in various ways into and from German and French and English:

> A first precaution against the risks of substantialism or irrationalism is to recall the differential character of force. In the texts I just evoked, it is always a matter of differential force, of difference as difference of force, of force as *différance* or force of *différance* (*différance* is a force *différée – différante*); it is always a matter of the relation between force and form, between force and signification, of 'performative' force, illocutionary or perlocutionary force, of persuasive force and of rhetoric, of affirmation of signature, but also and above all, of all the paradoxical situations in which the greatest force and the greatest weakness strangely exchange places.
>
> (D: 2002, AR pp. 234–35)

This reinscription of *différance* within the possibilities for analysis opened out by distinguishing between the neutral locution and its illocutionary and perlocutionary force underlines the manner in which writing is not simply a mater of what is written on a page, but a matter of how texts work. The analysis remaining concerned with the ambiguities of mapping distinctions in natural languages onto each other, opens the way to, but does not make the further move, made by Benjamin, of refusing altogether the temptation of a humanisation of time, in a concept of history as centred on human developments.

The third and latest address concerning Benjamin is '*Fichu*: Frankfurt Address', delivered on 22 September 2001, just after the 9/11 events. A footnote indicates that it opened in German: 'I apologize. I am getting ready to greet you and thank you in my language.' It continues:

> And language will be my subject: the language of the other, the visitor's language, the foreigner's language, even the immigrant's, the émigré's or the exile's. What will a responsible politics make of the plural and the singular, starting with the differences between the languages in the Europe of the future, and, as with Europe, in the ongoing process of globalization? In what we call, ever more questionably, globalization, we in fact find ourselves on the verge of wars that, since September 11, are less sure than ever of their language, their meaning and their name.
>
> (D: 2001, PM p. 164)

This address indicates quite plainly, if not directly, a lack of sympathy with Benjamin's dream of a language of pure reference, and a preference for Adorno, because Adorno seeks to minimise the role in Benjamin's thinking of political theology. Derrida invokes a mediating instance between himself and Adorno, who is also a mediating instance between Adorno and Benjamin:

> If Gretel Adorno were still alive, I would write her a confidential letter about the relationship between Teddie and Detlef. I would ask her why Benjamin doesn't have a prize, and I would share my hypotheses on this subject with her.
>
> (D: 2001, PM pp. 177–78)

Such mediation and such private communications form an undercurrent out of which public texts and public presentations emerge on to the surface for attention. The use of the diminutive form of Adorno's first name, Theodore, and of one of Benjamin's *noms de plume*, indicate a certain insider status for the speaker. Derrida thus indicates that he does not know where to place himself in relation to the difficult relation between Benjamin and Adorno.

That relation has been subjected to much discussion and analysis, concerned both with Adorno's borrowings from Benjamin, his role as Benjamin's literary executor, and his role in playing down the apocalyptic elements in Benjamin's writings. In this last regard, Adorno's relation to Benjamin might be likened to that of Derrida to Levinas. In the course of the lecture, Derrida does directly remark Adorno's lack of sympathy for Benjamin's use of a technique, in the construction of dialectical images, through which Benjamin analyses a conversion of the dream images of the nineteenth century into the realities of the twentieth century.[26] In these dialectical images, those past lost moments are briefly retrieved, on the model of a transitory redemption, affirmed in the shock of a moment of transformatory insight, and then lost again. For Benjamin, these moments fracture time, interrupting a continuum of an eternal recurrence, in which even the long dead are not safe from present persecution. The parallel here with Husserl's analyses of a cycle of fulfilling intuition and recurrent re-sedimentation remains to be explored. Nineteenth-century dream figures become the nightmare reality of the twentieth century, while the 'Concept 9/11' reveals a form of dating through which the politics of the next century has been marked out, breaking with the politics of nation and national determination, into new forms of extra-territorial political struggle. I shall pursue this notion of the historical event which breaks the continuity of history in the next part of these enquiries.[27]

Part V

Animal/machine

5.1 Cultivating limits

> Given that we find something purpose-like in nature's product, let us call nature's procedure (causality) a technic, and let us then divide this technic into an intentional and an unintentional one (*technica intentionalis* and *technica naturalis*). By an intentional technic, I mean that nature's ability to produce [things] in terms of final causes must be considered a special kind of causality; by an unintentional technic, I mean that this ability is quite identical with the mechanism of nature, and that we have falsely interpreted the contingent agreement of that ability with our concepts and rules of art, namely, as a special kind of natural production, whereas in fact, it is merely [the result of] a subjective condition under which we judge that ability.
>
> (CJ 72, AK V pp. 390–91)[1]

This citation from Kant's *Critique of Judgment* sets out a decision to be made, concerning the production of items in accordance with a perspective on the end result, between thinking this either as a mechanism of nature, compatible with a natural causality, or as needing a specification of a new kind of causality, through which the effects of human freedom or indeed the intendings of a divine being may be brought into the analysis. The decision here is constrained by Kant's presumption that the causality of nature precludes registering these effects of either human freedom or divine intending, or indeed of Intelligent Design, since he has an account of natural causality which presumes a greater degree of homogeneity amongst the forces and phenomena to be brought into the analysis, than more contemporary accounts of causes in nature might be inclined to do. Leaving the possibly anachronistic aspects of Kant's notion of causality in nature on one side, the question remains whether there is a difference, as marked up by Aristotle, between efficient and final causes, such that two distinct forms of linked processes are to be traced out when historical events and human activity are the focus for concern. The question thus posed reveals the further puzzle about the loss from the tradition of the notions of formal and material causes, as distinct from the final and efficient causes, which roughly speaking map on to the two forms of analysis contrasted here by Kant. When Derrida, and after him Bernard Stiegler, insist on an originary prosthesis, as simultaneous with an originary inception, and providing a necessary supplement for the analysis of its development, they in effect mark up a splitting between a material cause, and its efficacy, whereby materiality has implications above and beyond those taken up in the articulations of efficient causality.[2] The linkage presumed between formal and final causes can be similarly syncopated by insisting on delay and on non-simultaneities in the articulation of formal causes, such that they do not arrive at a single final realisation.

The terms of such a decision, thus constrained, need clarification with respect to an inheritance of a more elaborate account of causality, already available in Aristotle's writings in the *Physics*, in *De Anima*, and in *De Motu Animalium*.[3] This is then emphatically rethought by Heidegger, and complicated by the development of natural science, with the resulting transformation of concepts of technicity and technology.[4] The juxtaposition between mechanisms of nature and the artificial constructs of human ingenuity invites a further complication with the thought that the effects of human invention are not all intended results. Thus the notion of the prosthesis not only complicates the notion of a self-contained mechanism of nature, it also brings pressure to bear on any supposed transparency of human intending.[5] The analysis of writing in *Of Grammatology* gives an account of disseminations of meaning working above and beyond the specific intendings of individual human beings. The congruence of this analysis with aspects of Husserl's account of meaning lies in the analysis provided by Husserl both of the likely failure of meaning intending, to arrive at fulfilment, when the world turns out to be other than the meaning intending supposes, and in the failure to transmit ideal meanings, which have once found fulfilment, but which for whatever reason fall into the sedimented mode, lacking reactivation. Thus once the Kantian account of ends and means is complicated by an analysis of meaning as itself inscribed within systems of partially determinate teleologies of fulfilment and disappointment, the choice between an account in terms of mechanisms of nature and one in terms of an alternate form of causality, to permit analysis of systems of artifice and production, is complicated by a problem of distinguishing between natural and artificial systems at all. Focusing on language brings the problem to the fore, since language is pre-eminently a medium in which such a distinction falls into difficulty. For natural languages are natural by courtesy alone, and, in parallel with Heidegger's remarks about the essence of technology as not itself technical, the 'naturalness' of natural language is not itself natural.

The focus for attention in this section is the conjunction of the lengthy essays by Derrida, deriving from conference presentations in 1997 and 1998, 'The Animal Which Therefore I Am (More to Follow)' (*L'animal autobiographique*), and 'Typewriter Ribbon: Limited Ink (2)', from the conference on Paul de Man.[6] The conjunction of these two permits a positioning of the remarkable claim, in the now published longer version of '*L'animal autobio-graphique*' in *L'animal que donc je suis* (2006), about a link back to the transcendental unity of apperception, as hypothesised by Immanuel Kant, in the *Critique of Pure Reason*.[7] Derrida remarks that the Cartesian *cogito* and the Kantian account of the 'I think', which accompanies all my representations, play an important role in distinguishing human beings, among animals. Derrida discusses this 'I think', and the originary unity of apperception, in the context of its reprise in Kant's *Anthropology from a Pragmatic Point of View* (1798), where the human is identified as that which can have the 'I' in its representations.[8] Derrida then remarks:

> This power of having the 'I' takes on a dignity, it sets up, it erects the human infinitely above all the other animals which live on earth. This infinite elevation identifies a subject, in the strict sense, in so far as Kant underlines immediately after, that the 'I' indicates the unity of consciousness, which remains the same in all its modifications. The 'I' is the 'I think', the originary unity of apperception which accompanies all representation.
>
> (D: 2006, AQD pp. 129–30, my trans.)

Derrida then proceeds to ask a series of questions about a unique capacity for self-reference and self-indication, in which the role of naming and of being named by the other, of indicating and being indicated by the other, is complicated by the description of the experience of being looked at by his cat, a cat whose lineage goes by way of a reading of Lewis Carroll and Michel de Montaigne.

Derrida rehearses doubts about the adequacy of reasons given for privileging the human over other forms of animality. These are rehearsed as turning on some account of capacities distinctive of human beings, as demonstrated in language use, and reasoning, although Derrida also adduces the human predeliction for wearing clothes, such that human beings alone can be naked. Derrida analyses the manner in which a concept of life and of consciousness plays a role in securing this distinctiveness, which he then complicates in terms of connections between notions of life, and of the living present, and a notion of autobiography, and of confessional writing, which he has previously traced in their emergence in Husserl's writings, and in those of Augustine, and Jean-Jacques Rousseau. Later, in this 2006 text, drawing his remarks together Derrida adduces the following aside:

> At the heart of all these difficulties there is always the unthought of the thought of life (and it is here, with the question of life and of the 'living present', of the autobiography of the ego in its living present, that my deconstructive reading of Husserl began, and in truth all of that which could emerge from it).
>
> (D: 2006, AQD p. 153, my trans.)

Derrida suggests an enquiry into the unconscious of the Cartesian cogito and of the Kantian 'I think'. This 'unconscious' is the layer of the other in me, which thinks about me, and to which I respond, or follow on from. A discussion of this will here lead into a formulation of two contrasting sets of laws of composition for philosophical enquiry, at which point these enquiries arrive at a temporary standstill. There are the familiar laws of deduction, of parts and wholes, cohering into lines of argument and unifiable systems, with causal linkages for relations between items arranged in discrete domains of natural entities. Derrida, however, doubts whether parts and wholes can be taken to cohere into a unifiable system. He also doubts that there is a discrete domain circumscribing all and only natural phenomena. By contrast, Derrida's preferred operations conjoin a law of the symptom, borrowed from Freud, with a law of the three plus and minus one, borrowed from Hegel, as read through Jean Genet, whereby the priorities of quantity over quality, of wholes and parts, and of nature over prostheses are disrupted and dislocated, and in the very notion of a 'prosthesis' an ambiguity about a materiality or an ideality of any first term is displayed.

The paper entitled 'Typewriter Ribbon: Limited Ink (2)' was first presented at a conference on Paul de Man and the After Life of Theory. It recalls Derrida's exploration of de Man's writings, as formed by the notions of mourning and of promising, explored earlier in his *Memoires: For Paul De Man* (1986, 1989), and the issues opened up in the essay 'Limited Inc.' (1977), in which Derrida responded to John Searle's critique of his reading of J. L. Austin, in 'Signature, Event, Context' (1971).[9] The conjunction of *Memoires* and 'Limited Inc.' highlights the controversial reception of Derrida's writings in the United States. The conjunction of 'Typewriter Ribbon: Limited Ink (2)' and 'The Animal That therefore I Am' permits a reversal of the hypothesis, advanced in the previous section, concerning a conception of time, freed from its more customary embedding in a concept of history. The result is a conception of history, freed from an articulation of time. This is proposed by de Man, for reasons deriving from his logic of reading, but it reprises the structure of the historical *apriori*, hypothesised by Husserl, for a quite different set of reasons. Derrida thus rediscovers this structure, through a reading of de Man, and through a return to the emergence of a notion of genesis. This notion of genesis is triply determined, in his readings of Husserl, in his readings of Augustine's *Confessions*, and in a return to the ambiguities of the text, *Genesis*, with which both Jewish and Christian Holy Scriptures begin. This last reading confirms the thought that the theological-political events which constitute both Judaism and Christianity have no obvious location, either separately or together, in either a naturally given time series, or in a pre-existing historical narrative, or in a single act of divine creation, for the same story describes the creativity of the Christian and the Jewish God.

The earlier reading of de Man inflected Derrida's responses to Benjamin, and the differences between Derrida and Benjamin are perhaps best approached through Derrida's responses to de Man's readings of both Benjamin and Kant. Where de Man opens out a gap between time and history, affirming the autonomy of history, in order to permit a judgment of taste to occur, Benjamin opens out a gap between time and history to affirm a thinking of time, freed from the transcendental illusion of history, understood as progress, to permit a politics to take shape. As analysed in previous chapters of my enquiry, Derrida is committed to leaving the difference between such alternatives undecided, marking the possible arrival of an affirmative moment of self-inscription within that space of suspended judgment. Thus the disagreement between Benjamin and de Man is framed by Derrida's redeployment of the Husserlian gesture of bracketing. The resulting separation of time from history, and of history from time releases narrative and autobiography from any fixed position, serving an articulation of time as history. This confirms the gap between Derrida's construal of concepts and that of Paul Ricoeur, which ties a thinking of time to an analysis of narrative, and which ties an analysis of narrative to a subordination of figural language to a notion of literal or veridical use, to which some privilege is given.[10] Derrida pays attention to the notion of timeless structure and to the notion of an omni-temporality of essences, proposed by Husserl, as a counter to the emphasis on narrative, as attention to literary figurality works as a counter to the moves of all inclusiveness deployed in Hegelian dialectics.

Derrida disputes the legitimacy of either privilege, that of Hegel to dialectics or that of Ricoeur to narrative, and he affirms instead a practice of *circonfession*, which leaves the narrative of the other intact, but circumscribes it within a process of writing, into which the meanings of that previous narrative arrive. This is performed most explicitly in the text added to the 'Derridabase' provided by Geoffrey Bennington, the two together forming the text of *Jacques Derrida* by Geoffrey Bennington and Jacques Derrida (1991).[11] In this *Circonfession*, Derrida returns to a reading of Augustine's *Confessions*, and addresses himself to his own biography more directly than on any previous occasion.[12] In 'Derridabase', Bennington identifies a textual machine, or 'hypertext'program,

> which would allow at least in principle an almost instantaneous access to any page or word or mark from any other, and which would be plugged into a memory containing all of Derrida's texts, themselves simultaneously accessible by 'themes', key words, references, turns of 'style', etc.
>
> (B&D pp. 314–15)

He continues:

> Such a textual machine would not in the last instance be a pedagogical, technical tool, nor an efficient and technologist way of 'learning Derrida', although it is undeniable and not at all regrettable that it would also lend itself to such uses – for the program would also include instructions displaceable according to a chance that would exceed any programming mastery by opening that mastery to it.
>
> (B&D p. 315)

This notion of a textual machine is one which both invokes the discussion by Kant in the *Critique of Judgment*, of a difference between intentional and non-intentional technics, and undercuts the distinction between a technics of nature, understood as mechanism, and a notion of final causality, understood as separate and separable from a domain of nature.

While the practice of *circonfession* is deployed most obviously in this marginal text wrapped around the account of Derrida's enquiries written by Geoff Bennington, the technique had already been used in '*Survivre – journal de bord*', or 'Living On – Border Lines', to put in question the possibility of translation even between languages as cognate as French and English.[13] This marginal writing simultaneously suggests an overlap between the concerns of translation, for literary texts, and the process of transference, *Übertragung*, as a central feature of psychoanalytical technique. It also twists together the practices of translation and of literary critique, with a third strand, of self-inscription, signature and autobiography. These linkages come to the fore already in the 1970s in a series of writings on Nietzsche, some published in *The Ear of the Other*, in which the notion of *otobiography*, literally, writing the life of the ear, of listening, is coined.[14] This marks up the importance for Nietzsche, of the ability to hear the significance of certain distinctions, tones or shifts of register. There is then a distinct resonance between the inscription of a relation between life and thought in Nietzsche's writings, and the enterprise of deconstructive critique, in which the task of responding to the precarious unity of a critical consciousness is taken up. Nietzsche's ears join Husserl's silent speech and Heidegger's hand to pose a question to the disembodied state of Dasein, as marked up for discussion in Derrida's readings of Heidegger, in the series of papers entitled *Geschlecht*, from the eighties.[15] Derrida's analyses of Heidegger's hand, in *Geschlecht* II, and of the ear of the other, in *Geschlecht* IV, take up and develop Levinas' question: does Dasein eat? This points up how, because the analysis in *Being and Time* is left open-ended, the connection between the fundamental ontology of Dasein, and the empirical existence of human beings is left undecided.

Dismemberment of the human body, as a figure for an unattainable unity of an organic meaning, is marked up by de Man in his reading of Kant's Third Critique, published posthumously in 'Phenomenality and Materiality in Kant', in *Aesthetic Ideology*.[16] In another essay printed in this text, 'Kant's Materialism', de Man disputes any dematerialisation of Kant's aesthetics. He reads Kant's text to expose a disjoining of materiality at work in its textual articulation. Derrida's readings of, and responses to, Kant are thus to be conjoined to de Man's challenge to an ideology which seeks to locate aesthetics within idealism, as opposed to discovering within some of the classic texts, even those of Kant and Hegel, an analysis of the material conditions for aesthetic judgment and of their material media of articulation, which subverts any easy classification of them, as irreducibly privileging an idealism over materialism. De Man's readings of Kant and of Hegel thus adjoin Husserl's construal of transcendental aesthetics and of a transcendental idealism, as neutral between idealist and materialist interpretations. While there is much agreement between de Man and Derrida, Derrida's return to a reading of de Man is marked by a painful history. From their meeting in 1966 at the conference at Johns Hopkins, at which Derrida delivered 'Structure, Sign and Play in the Human Sciences', theirs had been a close, if mutually challenging relation.[17] To the second edition of the series of lectures given at Yale, *Memoires: For Paul de Man*, on memory and mourning, marking the death of his friend, Derrida added the essay 'Like the Sound of the Sea Deep within a Shell: Paul de Man's War'. This responds to the shock of the posthumous discovery of de Man's proto-fascist essays, published as a newspaper column in occupied Belgium, from 1940 to 1942.[18] Derrida's return to this relation in 1997 is marked by a re-opening of the relation to Kant and to Benjamin, through a questioning of self-constitution, and as articulated through the response to this especially contentious other.

The living present conjured up by the conjuncture of these two essays 'Typewriter Ribbon: Limited Ink (2)' from the conference in 1998, and 'The Animal Which Therefore I Am', from the 1997 conference on 'The Autobiographical Animal', is one of retrieval, conjunction and disjunction. At the beginning of the latter, Derrida marks up the two previous occasions on which conferences had been held at Cérisy-la-Salle, focused on his work. To the first of these, 'The Ends of Man', he delivered 'Of an Apocalyptic Tone Recently Adopted in Philosophy' (1980), which addresses the unstable tone of innovative and politically committed writing, and responds to Kant's essay, 'On a Newly Arisen Superior Tone in Philosophy' (1796). At the second, 'The Passage to the Frontiers' (1992), he delivered the lecture which forms the basis of *Aporias: Dying – Awaiting (One Another at) the 'Limits of Truth'* (1993). In each, a relation to Kant is staged, through the invocation of Kant's title in the first address, and in the notion of 'limits' from the First Critique in the second. The importance of Kant for Derrida becomes overwhelmingly clear in the texts, translated in *Who's Afraid of Philosophy: Right to Philosophy 1*, and *Eyes of the University: Right to Philosophy 2*.[19] This connection back to Kant underpins a tracing of his own intellectual trajectory, and his own preoccupation with the questions: who is taught philosophy, and what is the connection between philosophy and legal entitlement, which are pursued across the various papers published in these two volumes. In this third conference address, Derrida interweaves a theoretical philosophical strand, and a biographical strand, a questioning of institutions and of political entitlement.

While Heidegger, Blanchot, Levinas and Derrida can be seen to be in dispute for the inheritance of Husserl; Ricoeur, Benjamin, de Man and Derrida can be seen to be in dispute for the inheritance of Kant, and both lineages are in dispute with any affirmation of the strong Hegelian programme, of systematically linking the results of the disciplines of philosophy, art and religion. Each dispute takes the form of the mode of counting, developed by Derrida in opposition to Hegel: three, plus or minus one. By conjoining a question of the inheritance of Kant's critical enquiry, to that of the inheritance of the phenomenology of Husserl, there opens out a contestation between Kant and Husserl about how to interpret the connections between time, space and the formation of concepts. In conclusion to these readings I suggest then that Derrida's notion of *différance*, and its various rewritings, as *destinerrance*, otobiography, topolitology and limitrophy, are all responses to the question: how to think a transcendental aesthetics, which accepts Husserl's move of undercutting the distinctions constitutive of Kant's critical enquiries: between the faculties of intuition, understanding and reason, and between an *apriorism* of sensation, in the forms of intuition, and an *apriorism* of understanding, in the categories of the understanding. Husserl identifies a more basic form of *apriorism*, deriving from determinations of meaning, in primordial meaning intuitions, which themselves bear empirical historical indices. The basis on which these transformations can be thought is that of a set of syntheses, as hypothesised by Husserl, but taking place, each time uniquely, as a delimitation of a world within the world, which is the gesture of autobiography. This then takes the place of Kant's positing of a transcendental unity of apperception, as the site at which the various faculties come to bear on one another, to demonstrate how meaning and knowledge are possible. Autobiography, as *circonfession*, self-inscription and the signature take the place of a transcendental unity of apperception, as the basic structure holding together articulations of language as meaningful: as a transcendental unity of auto-inscription.

At the third conference, Derrida delivered, as a series of lectures, the parts of *L'animal que donc je suis*. In its various parts, a history of failing to think the animality of the human, from Descartes on, is set out, gesturing forwards to works as yet to arrive in the public space, just as 'Typewriter Ribbon' gestures back to works already in the public space, but not yet placed in a trajectory of enquiry, for which the possibility of, and conditions for, self-ascription are indispensable. For such self-ascription, connections between names, naming, naming oneself, signing, autobiography and the condition of mourning are all important. In 'The Animal Which Therefore I Am', Derrida writes of the biblical *Genesis* narration, rehearsing his now famous procedure of positing an alternation between 'on the one hand' and 'on the other hand':

> On the one hand, the naming of the animals is performed at one and the same time, before the creation of Ishah, the female part of man, and as a result before they perceive themselves to be naked; and they are at first naked without shame . . .
>
> On the other hand, and this is especially important, the public announcing of names remains at one and the same time free and overseen, under surveillance, under the gaze of Jehovah, who does not for all that intervene. He lets Adam, he lets man, man alone, Ish without Ishah, the woman, freely call out the names. He lets him go about naming alone.
>
> (D: 1997, AA p. 385)

On the following page, Derrida then writes:

> God lets Ish call the other living things all on his own, give them their names in his own name, these animals that are older and younger than him, these living things that come into the world before him but were named after him, on his initiative according to the second narrative. In both cases, man is, in both senses of the word, after the animal. He follows him. This 'after' that determines a sequence, a consequence or persecution, is not in time, nor is it temporal; it is the very genesis of time.
>
> (D: 1997, AA p. 386)

Here there is, first, a time, in which God creates, first, man, and then woman. There is, then, temporal sequence, the order of exposition and the order of these animals both older, created first, and younger, named second. In the third place, but as condition of both, there is 'the very genesis of time', thus performing a sequence of reversals, whereby the first and the second turn out to be conditional on something in advance of the first.

It is not insignificant that this reversal is tied to the ambiguity of the generation of sexual difference. For in this account of creation in *Genesis*, God creates first man, and then man names the animals, and then God creates woman, setting up one form of generativity; however, in the first account in *Genesis* of the creation of humanity, God creates man and woman simultaneously on the fifth day, 'male and female created he them'. On the fourth day, God creates the sea creatures and the birds, and to these animals of the sky and the sea, he says: 'Be fruitful and increase; fill the water of the sea and let the birds increase the land' (*Genesis* 1.21). On the fifth day,

> 1.24: God said, 'Let the earth bring forth living creatures, according to their various kinds: cattle, creeping things and wild animals, all according to their various kinds.' So it was;
>
> 25: God made wild animals, cattle and every creeping thing, all according to their various kinds and he saw that it was good.
>
> 26: Then God said 'Let us make human beings in our image, after our likeness to have dominion over the fish in the sea, the birds in the air, the cattle, all wild animals on land, and everything that creeps on the earth.'
>
> 27: God created human beings in his own image; in the image of God he created them; male and female he created them.

This account concludes: '2.3: God blessed the day and made it holy because it was the day he finished all his work of creation.' Derrida points out how the account begins again, with 'This is the story of the heavens and the earth after their creation' (*Genesis* 2.4). This structure of reversal, in relation to the genesis of time, and of an understanding of temporality can be encountered also in the reconstruction of Husserl's thinking about time. For Husserl, of course, the *apriori* is, like God, there from before the beginning of time, and the genesis of time must be thought of as preceding the beginning of time.

In this second story, there is, as Derrida remarks, 'at least a type of new beginning, a second beginning which is distinguished in "some translations" as the second narrative' (D: 1997, AA p. 383). This 'second beginning' is to be heard as an echo of the 'other' beginning, as invoked by Heidegger, in *Of the Event: Contributions to Philosophy* (1989). The emphasis here is not on God's creation as work, nor yet on the performative utterance, whereby speech is creation, but on a time lapse between already existing entities, to which names are retrospectively assigned. 'So from the earth he formed all the wild animals and all the birds of the air, and brought them to man to see what he would call them;

whatever the man called each living creature, that would be its name' (*Genesis* 3.16). Eve too is named by 'man', retrospectively, at 3.20, while 'man' does not name himself, but is named Adam, from the Hebrew *adamah* for earth: '2.7: The Lord God formed a human being from the dust of the ground and breathed into his nostrils the breath of life, so that he became a living creature.' This account of the human, in advance of sexual difference, as a combination of 'ground' and 'the breath of life', air and earth, sets up a different play of differences, out of which a distinctive, elemental generativity may be configured. There are thus here two articulations of generativity and, intriguingly, two distinct processes of naming, one of which proceeds forwards, a divine naming calling humanity into existence, and one of which proceeds backwards, retrospectively, once humanity has been distinguished into two distinct sexes, naming the animals after they have been created. The collective term, *adamah*, is then appropriated to one side of sexual difference, and the subordinated half is called 'Eve'. Then the supposedly subordinated supplementary half, Eve, defies this subordination and tempts the first named to eat of the apple of knowledge. Thus the order and the hierarchies are stabilised and destabilised.

The theme of a genesis of time emerges out of a prior process of naming, in this case a process of naming heavily over-determined by invocations of a time before sexual difference, on the possibility of which Derrida, in relation to Dasein, as divided and collected under the term *Geschlecht*, casts doubt. As Derrida observes, this is a time before original sin, where the one who accompanies me is a god, split, as Derrida observes, between 'an all powerful God, and the finitude of a God who doesn't know what is going to happen to him with language. And with names.' This unknowing grounds, perhaps, another difference; for the language of God creates, as it names, a language of simultaneous fulfilment. The language of finitude, by contrast, articulates a time lapse between creation and a name, with a fulfilment which can only follow on after, and which is subject to delay and distortion. The determinations of time in this text, in which Derrida evokes both *Genesis* and genesis, as in generativity, are multiple. There is the time of duration, from 1997, the time of the delivery of 'The Animal Which Therefore I Am', back to 1959, the time of Derrida's first public lecture, on 'Structure and Genesis', placing Husserl's conception of genesis in the context of the then current notion of structure. Of this Derrida remarks:

> If already I were to give in to what others might call the instinct of the autobiographical animal, I might recall that in 1959, as today, the theme was, in short, Genesis. The title of the conference was 'Structure and Genesis' and it was my first ten day Cérisy event.
>
> (D: 1997, AA pp. 370–71)

This then poses a question to changes of context, in the reception of Husserl, as marked in 1959 by structuralist analysis, as in conflict with the claims of phenomenology, as marked in the 1980s, by Jean-Luc Marion's deployment of phenomenology, to affirm a religious commitment to a real presence in the eucharist. Marion's affirmation of Husserl's phenomenology is not obviously compatible with it, nor is structuralism so obviously incompatible with it.[20]

This time of succession suggests that 1997 should be thought to follow 1993, and 1980, and 1959, with a certain irreversibility; while by contrast, for a time of the symptom, of reversibility, or of delayed affect, the later text can be read to elucidate the earlier. The textual determinations in the first three chapters of *Genesis* are also multiple. There is the time of a duration, seven days of creation, the duration of God's work, which presumably contains within it all times; and the time of a succession, of day after day. There is a time of animals, with a delay between their creation, and their naming, and then there is the time of the language of tenses, conditionalities and reversibilities. There is, on one side, a time before time, the time of genesis, the genesis of time; and then there is a description of it, in a book, *Genesis*, which describes in a certain language, Hebrew, with a certain syntax and semantics, the genesis of time and order, and the emergence of humanity in the order of divine creation. There is the time of repetition, or supplementation, of the double narration of the creation of heaven and earth, and there is the double telling of the creation of human beings, and of their generativity. There is then finally an unhingeing of divine and human time, and of divine and human language, which cannot be made commensurable with one another.

Derrida, at the start of the 1997 address, invokes Nietzsche's observation from *Genealogy of Morals*, that human beings are the species of animal which can make promises. This entails that this animal is also capable of misleading speech (*das Versprechen darf*) and of lying.[21] Freud returns to this theme, giving it an added emphatic twist, since for him conscious speech is always at the mercy of *parapraxis*, of an unconscious slippage, known as the Freudian slip, in which a concealed meaning is by chance substituted for an avowed meaning. The Freudian logic here licenses the very Derridean move of suggesting that a straightforward opposition between promising and veracity, on one side, and lying and falsehood, on the other, is disrupted in advance by a third force, the chance, or necessity of an uncontrolled speech, in which some other order of meaning is revealed. Husserl, too, has this three-part structure, in his analyses of meaning, with *Sinn*, meaning, contrasted to both *Unsinn*, meaninglessness, and *Widersinn*, absurdity or contra-sense, which appears syntactically sound, but which results from failed attempts to deploy language correctly. De Man in his reading of Benjamin draws attention to a further splitting, for Husserl, between the referentiality of *Bedeutung*, and *Bedeuten*; the assigning of meaning, and the terms through which distinctions between them can be arrived at, through analyses of *Meinen*, sometimes translated as opining, what is actually thought, and in *die Art des Meinens*, the manner in which meanings come available.[22] This second and third series of differentiations mark up differences between attempts at objective reference, and subjective opining, in the process of seeking to arrive at a determination of a pure order of *Sinn*.

Husserl's notion of a pure sense, as neutral with respect to natural languages, returns to haunt Derrida, in the difficulties posed for him by Benjamin's invocation of a 'language as such' which, for Benjamin, is indicated by the points of contact between a translation and that which it attempts to translate. This tangentiality is distinct from that to be rehearsed by Derrida in relation to Jean-Luc Nancy.[23] Derrida remarks how, while promising is allied to speech, and the possibilities of truth and falsehood, Benjamin marks an alliance of muteness to mourning:

> It is true that according to Benjamin, the sadness, mourning, melancholy (*Traurigkeit*) of nature and animality are born out of this muteness but also out of and by means of the wound without a name: that of having been given a name. Finding oneself deprived of language, one loses the power to name, to name oneself, indeed to respond to one's name.
>
> (D: 1997, AA p. 388)

The Benjamin text, cited here by Derrida, 'On Language as such and the Languages of Human Beings' (1916), he carefully locates in its time of writing, in the midst of the First World War.[24] Derrida then adduces a remark of Heidegger's from *Being and Time* (1927), some eleven years later, oddly noting neither the time lapse, nor Benjamin's pronounced hostility to Heidegger, nor indeed referring back to his own previous encounters with the writings of Benjamin, not all of them well augured.[25] Language as such, like the pure grammar, and the *eidos* and *eide* of Husserl's analyses, remains mute, and only the languages of humanity and their imperfect corrigible meaning intendings articulated in natural languages make themselves heard as speech. Language as such is the medium for naming the death in life, a saturated meaning, which therefore cannot fall into the natural time series. Husserl, Benjamin and indeed Marion attempt to make sense of such muteness, as bearing meaning, an attempt which for Derrida has the status of absurdity. The '*einsame Seelenleben*' the solitary life of the spirit, and the '*monologe Rede*', discursive monologue, of Husserl's first *Logical Investigation* are implicitly invoked here.

The remark cited by Derrida from Heidegger's *Being and Time* on animals reveals that, for all his emphasis on the priority of Dasein with respect to the question of being, its forgetting and the erasure of the question of time, the animal too is constituted by some kind of time. The status of animality provides an instance of some kind of 'in between', between a timelessness or omnitemporality of numbers and the self-constituting temporality of Dasein:

> Only an entity, which in accordance with the meaning of its Being, finds itself in a state of mind – that is to say, an entity, which in existing is as already having been, and which exists in a constant mode of what has been (*Gewesenheit*) – can become affected. Ontologically such affection presupposes making-present (*Gegenwärtigen*), and indeed in such a manner that in its making-present, Dasein can be brought back to itself as something that has been. To define ontologically the way in which the senses can be stimulated or touched in something that merely has life (*in einem nur Lebenden*)

and how and where the Being of animals, for instance is constituted by some kind of 'time' remains a problem for it (*bleibt ein Problem für sich*).

(H: SZ 346/MR p. 396 trans. mod.)

That which only has life has no access to the crucial determination, for Heidegger, of being-towards-death. The temporality of this affectivity, where the senses are stimulated in something 'that merely has life', cannot be determined, because it lacks delimitation. What for Heidegger distinguishes Dasein from the animal is that Dasein has a relation to its own limitation in death, in the mode of *Sein zum Tode*, being-towards-death, which articulates the present moment as open to the other temporality of another kind of time, in a future of no longer being alive. The human as animal perishes, but, as Dasein, is determined as finite, with determinate limits, provided by its relation to being, to world, to meaning and to time. As no-longer Dasein, this human existence can be mourned and memorialised, constituting mourning as an ever present possible relation to others and indeed to oneself. Dasein itself is neither in time, nor subject to a before and after, but as the source of an articulation of time itself, it inaugurates another temporality.

This other temporality is invoked by Freud, under the rubric of the death drive, by contrast to the pleasure principle. The connections and disjunctions between these two thinkings of death provide the route into the reactivation of Hegelianism given by Derrida in *Glas: What Remains of Absolute Knowledge* (1974), as not delimited by Hegel's dialectical closure. Derrida remarks this 'life in its simple state', and adds

> I will also ask myself whether this fiction, this simulacrum, this myth, this legend, this phantasm of what is offered as a pure concept (life in its pure state – Benjamin also has confidence in what can probably be no more than a pseudo-concept) is not precisely pure philosophy become a symptom of the history which concerns us here. Isn't that history the one that man tells himself, the history of the philosophical animal, of the animal for the man-philosopher?
>
> (D: 1997, AA p. 391)

As a response to this interrogation, silencing the violent conjugation of the names 'Heidegger and Benjamin', Derrida rehearses his two hypotheses concerning time:

> *First hypothesis*: for about two centuries, intensely and by means of an alarming rate of acceleration, for we no longer even have a clock or a chronological measure of it, we, we who call ourselves men or humans, we, who recognize ourselves in that name, have been involved in an unprecedented transformation. This mutation affects the experience of what we continue to call imperturbably, as if there were nothing wrong with it, the animal and/or animals. I intend to stake a lot on the flexible separation of

> this *and/or.* We must continuously move along this coming and going between the oldest and what comes of the exchange among the new, the 'again' and 'anew' of a repetition. Far from appearing, simply, within what we continue to call the world, history, life and so on, this unheard of relation to the animal or to animals is so new that it should oblige us to worry about all those concepts, more than just problematize them.
>
> (D: 1997, AA pp. 392–93)

The oldest is this naming of the animals, and a naming of Eve, by Adam under the watchful gaze of an unreconstructed divinity. This 'again and anew' of a repetition permits Derrida to surmise a theology of naming both with and without binding religious commitment.

Derrida then adduces two hesitations, because he supposes that a transformation in the relation between humans and other animals subordinates any 'living through' to a critical turning point:

> That is why I would hesitate to say that we are living through that (if one can still confidently call life the experience whose limits tremble at the border crossings between *bios* and *zoe,* the biological, zoological, and the anthropological, as between life and earth, life and technology, life and history and so on).
>
> (D: 1997, AA p. 393)

It is instructive to note that while in the former quotation, the list of concepts rather to be worried over than problematised is given as 'world, history, life and so on', the first hesitation concerns precisely life, the term marked up by Heidegger in his remark about animals, which connects back to Husserl's notion of transcendental life. There is a kind of living which all animals, including the human, experience, but there appears to be reserved a kind of life which is enjoyed only by the animal as human. This distinction between kinds of life is then connected to the capacity to deploy language, and to assign and bear names, but it remains radically underdetermined. Derrida continues, by announcing his second hesitation, taking up the second term in his list, history:

> I would therefore hesitate just as much to say that we are living through an historical turning point. The figure of the turning point implies a rupture or an instantaneous mutation for which the model or the figure remains genetic, biological, or zoological and which therefore remains, precisely, to be questioned. As for history, historicity, even historicality, those motifs belong precisely – as we shall see in detail – to this auto-definition, this auto-apprehension, this auto-situation of man or of the human Dasein with respect to what is living and with respect to animal life; they belong to this auto-biography of man that I wish to call into question today.
>
> (D: 1997, AA p. 393)

To this series of hesitations it may be necessary to mark up the unresolved hiatus concerning sexual difference, and the unnameable differences between the divine naming, which calls an entity into existence, and a human naming which names retrospectively, a naming marked by delay.

In this list 'world, history, life and so on', 'history' and 'life' have been discussed already; the concept 'world' has not yet received attention, although it is both critically in dispute between Kant, Heidegger and Husserl, and emphatically thematised by Derrida elsewhere, in 'Faith and Knowledge: The Two Sources of "Religion" at the Limits of Reason Alone', (1996) in terms of an unstable distinction between the Anglicism 'globalisation' and the Gallicism *mondialisation*, which Derrida points up in its connection to latinisation, as *mondialatinisation*.[26] Thus the classical phenomenological theme, concerning world and worldhood, is complicated by Derrida, who draws attention to the specific languages, in which the unity of the world, as geopolitical entity, is affirmed: the Latin of republic, empire, and church, and the English of British and American empire. This work on the concept of world performed respectively by Husserl and by Heidegger presupposes the gap opened up by Kant's critical philosophy between 'world' conceived of as transcendental limit, a 'world' conceived as sum of empirical circumstances, and a 'world' conceived as the sphere of human political ambition and of cosmopolitical intent. For there is a contrast for Kant between the cosmopolitics discussed in his late essay 'Idea for a Universal History Viewed from a Cosmopolitical Point of View', and a universal cosmology.[27] *Mondialatinisation* is to be contrasted to both Husserl's life world, in which a parallelism between empirical psychology, phenomenological psychology, and transcendental phenomenology is experienced; and to the worldhood Heidegger deduces, in what can be read as a phenomenological reduction of such experience of a life world. For Husserl, the concept of world is to be linked to a concept of the horizon within which phenomena appear as phenomena, and the task is to construct the most inclusive possible concept of world, such that the resulting horizon is all-inclusive. In *Being and Time*, Heidegger reduces this worldhood to the structure of Dasein's temporality, given as ideal meaning, thus generating an account of the priority of temporality over this worldhood. This, however, should still not be understood as a decision for an idealism and against a materiality, for a thinking in terms of an idealisation of meaning does not predetermine the ontological status of what is meant.[28] Such formalisation precisely leaves the ontological status of its fulfilments open for interpretation.

Later in the essay, Derrida evokes Heidegger's 1929–30 lectures, *Basic Concepts of Metaphysics: World, Finitude, Solitude* (1975), in which Heidegger writes that the animal is poor in world.[29] For Heidegger, while animals are in time, and articulate themselves in time, they do not, apparently, have access to the third level, called here the genesis of time, which both provides a thinking of time as delimited, and permits a thinking of a conjunction and disjunction of the concept of time, made available by being-in-time, with a concept of temporality, as the 'through which' of living. This 'through which' of living is delimited by the concepts of birth and death, which provide determinate limits, thus securing the finitude of Dasein. This Heideggerian temporality, achieved as a result of operating a second reduction on the notion of worldhood, leads into a re-articulation of the Husserlian notion of time, as genesis and opening, as *Zeitigung*. To this Derrida has a double objection, that it fails to do justice to the Hegelian rethinking of time, and that it fails to do justice to the forces of re-stabilisation which, Heidegger's protests to the contrary, will reimpose a delimited concept of time, in place of a thinking of time as opening. The key move to be contested is Heidegger's claim that the genesis of the temporality of Dasein is given in the first instance in the mode of Dasein's historicality: that the temporalising structure of temporality (*die Zeitigungsstruktur der Zeitlichkeit*) 'reveals itself as the historicality of Dasein' (H: SZ p. 332 / MR p. 381). Heidegger specifies primordial historising in the following way: 'This is how we designate (*damit bezeichnen wir*) Dasein's primordial historizing, which lies in authentic resoluteness and in which Dasein hands itself down to itself, free for death in a possibility which it has inherited, and yet has chosen' (H: SZ p. 384 / MR p. 435). Heidegger's account of time and history remains radically incomplete in *Being and Time*, and the change of focus in the subsequent work does not solve the difficulty. It is acceptable to Derrida neither in its incompleteness, nor in any hypothetical subsequent completion.

Derrida's second hypothesis follows a few pages later: 'Here now in view of another thesis is the second hypothesis that I think must be deduced without hesitation.' He continues:

> It concerns or puts into effect another logic of the limit. I will thus be tempted to inscribe the subject of this thesis in the series of the conferences that beginning with '*Les fins de l'homme*' and followed by '*Le passage des frontières*' have been devoted to a properly transgressal, if not transgressive, 'experience of *limitrophy*', that is the cultivation of limits and the limits of cultivation.

And it is perhaps worthy of remark that while Derrida says that this hypothesis is to be deduced without hesitation, he in fact then introduces a delay, in the list of contributions to the previous Cerisy conferences, providing the notion of limitrophy with a genealogy, in his previous analyses of the similarly transgressal notions of topolitology and of *destinerrance*. The former, as discussed in 'How to Avoid Speaking: Denials' (1984), and in my Part IV above, sets out a distinction between the human and the animal thus:

> The essence of such a secret would remain rigorously alien to every other non-manifestation; and, notably, unlike that of which the animal is capable. The manifestation or non-manifestation of *this* secret, in short, its possibility, would never be of the order of the symptom. An animal can neither choose to keep silent, nor keep a secret.
>
> (D: 1989, HA p. 87)[30]

Again the opposition, manifest/non-manifest, is disrupted by the logic of the symptom, which is neither the one nor the other. In 'The Animal Which Therefore I Am', Derrida continues, concerning a certain cultivation of limits, now to be called 'limitrophy':

> Let's allow this word to have both a general and strict sense: what abuts onto, limits, but also what feeds, is fed, is cared for, raised, and trained, what is cultivated on the edges of a limit. In the semantics of *trepho*, *trophe*, or *trophos*, we should be able to find everything we need to speak about what we should be speaking about in the course of these ten days devoted to the autobiographical animals: feedings, food, nursing, breeding offspring, education, care and keeping of animals, training, upbringing culture, living and allowing to live by giving to live, be fed, and grown autobiographically. Limitrophy is therefore my subject.
>
> (D: 1997, AA p. 397)

While the link back to Nietzsche's remarks on the breeding of animals remains implied and oblique, these senses, both general and strict, are replicated through the open-ended series of splittings: two hands, two hypotheses, two hesitations, which only by *fiat* remain fixed at the number 'two', rather than becoming three plus or minus one.

The 'general' sense concerns the work of artifice in the cultivation of delimitation, whereas the 'strict' sense concerns the care for the animal's growth and well-being, but also, presumably, for its death, at a suitable or unsuitable time. Thus topolitology, which provides an analysis of the experience of place, in relation to the name of God and its unsayability, is here to be displaced by an analysis of a propagation, which unsurprisingly will abut on to the delimitation, in Kantian critique, of the deployment of concepts; and on to the separation of agencies in the Freudian topographies. Derrida's questioning of time is thus nested in the juxtapositions of biblical and Husserlian genesis; and between Kantian and Freudian delimitations and topographies. Limitrophy follows on from topolitology, as topolitology follows on from otobiography, and otobiography from grammatology: each with the form of a scientificity, or specific disciplinarity, which brings into question the procedures for accepting and rejecting claims on scientificity. This questioning, like the questioning of how it is possible to read texts, has led to a failure of nerve in Derrida's readers and to vulgar denunciatory gestures. With these paradoxical and inherently unstable words, he seeks to train listeners to hear, to make audible and intelligible, problems concerning instabilities of meaning, which, on the limits of intelligibility, are barely audible. This limitrophy marks the inception of an account of an atrophying of time, in philosophical attempts to delimit and constrain within conceptual boundaries a contrapuntal syncopated flow.

Certainly Heidegger in *Being and Time* is of the view that philosophical accounts, in privileging a 'now' of perception and intuition, have erased a more basic notion of time, and indeed the genesis of both time and being. He gives an account of the genesis of 'the vulgar concept of time', in section 82, and he shows how it displaces and erases the traces of a more complicated, many-dimensional temporality, which is not fixed in a mode of presentation, in the present. 'We shall call the world time made visible in this way by the use of clocks the *Jetzt-zeit*' (H: SZ p. 421 / MR p. 474), Heidegger writes, with another gesture of adamic naming. He juxtaposes this *Jetzt-zeit* of the sequence of no-longer earlier and not-yet later moments, as the time of measuring movement, to the momentary *Augenblick*, an ecstatic unity of Dasein's self-identification as finite, stretched out between birth and death, and projecting into its past, its present concerns and its futurity as being-towards-death. In the mode of present concern, it falls into the restricted 'now' of calculation, which is in time, but is not time itself. The momentary *Augenblick*, the blink of an eye, is not in time: it is a rupture of time, and therefore this, too, is not time itself. It is perhaps worth remarking that this *Augenblick* is the word used for the biblical moment of divine judgment, when at the end of time the dead are resurrected, human beings are made whole, and enter the Kingdom. Heidegger seeks to detach the word from its biblical context, and to re-deploy it in relation to an account of history, as derived from a more basic historicality, now understood as that which makes it possible for Dasein to understand itself in relation to inheriting a past in the present, and as having a relation to a future, which it helps to bring about.

Heidegger remarks that Dasein occurs for the most part in the mode of inauthentic historicality:

> If historicality belongs to the Being of Dasein, then even inauthentic existing must be historical. What if it is Dasein's *inauthentic* historicality that has directed our questioning of the 'connectedness of life' and has blocked off our access to authentic historicality and its own peculiar 'connectedness'? However this may be treated, we cannot do without a study of Dasein's inauthentic historicality, if our exposition of the ontological problem of history is to be adequate and complete.
>
> (H: SZ p. 387 / MR p. 439)

Heidegger thus repeats the structure of grounding the authentic in the inauthentic, which he insists on when he first introduces the notion of the authentic, a relation of grounding which he derives from Husserl's account of meaning. Derrida's first discussion, in '*Ousia* and *Gramme*', in the 1960s, of the relation between the authentic and the inauthentic, links into his reading of Husserl, and the move into readings of both Heidegger and Hegel, in terms of a repeating return of a metaphysics of presence.[31] Derrida's return to the relation of grounding between authentic and inauthentic modes is under the escort of his reading of Blanchot and Levinas, in *Aporias*, at the second Cerisy conference on Derrida's work. In the first reading, he questions the distinction between authentic and inauthentic concepts of time; in the second reading he questions the distinction between authentic and inauthentic concepts of death. In dispute here is the possibility of moving from the arbitrary gesture of an individual,

Heidegger, naming an arbitrarily selected 'this', with an arbitrarily selected name, and a naming which picks out an ideal meaning, held in place by the nature of what there is: the difference between contingent generalisation, and a necessary formalisation, with a claim on atemporal truth.

In the latter part of *Being and Time*, Heidegger introduces an authentic historicality, by contrast to inauthentic historicality:

> In inauthentic historicality, on the other hand, the way in which fate has been primordially stretched along has been hidden. With the inconstancy of the they-self Dasein makes present its 'today'. In awaiting the next new thing, it has already forgotten the old one. The 'they' evades choice. Blind for possibilities, it cannot repeat what has been, but only retains and receives the 'actual' that is left over, the world-historical that has been, the leavings, and the information about them that is present-at-hand. Lost in the making present of the 'today', it understands the 'past' in terms of the 'Present'.
>
> (H: SZ p. 391 / MR p. 443)

He sets out the contrast:

> On the other hand, the temporality of authentic historicality, as the moment of vision of anticipatory repetition, *deprives* the 'today' of its character as *present* (*eine Entgegenwärtigung des Heute*) and weans one from the conventionalities of the 'they'. When, however, one's existence is inauthentically historical, it is loaded down with the legacy of a 'past' which has become unrecognisable and it seeks the modern. But when historicality is authentic, it understands history as the 'recurrence' of the possible, and knows that a possibility will recur only if existence is open for it fatefully, in a moment of vision, in resolute repetition.
>
> (H: SZ p. 391 / MR p. 443–44)

It should be noted that what is translated here as 'on the other hand' is, in German, the adverb '*dagegen*', which should rather be read alongside the 'da' of the term Dasein, than alongside Derrida's analyses of hands. Derrida again has a question to this strategy of derivation, of grounding inauthentic historicality in authentic historicality, as well as questioning the basis for opposing the two, which the detour through the reading of Husserl opens up. For, as remarked, it is Husserl who introduces these relations of derivation and ground, of grounding the inauthentic in the authentic mode.

A sequence of nows, unbroken, without gaps, and non-finite, is opposed to a time of breaks and transformations, which Heidegger seeks to ground in the finitude of Dasein, in *Being and Time*, and in the *Ereignis* of being, in the works of the turn, where he retrieves another biblical word, used for the analysis of the moment of conversion to Christianity, from his early lectures on phenomenology.[32] In the next section of *Being and Time*, he sets out a contrast between his own view of time, as grounded in finite Dasein, and Hegel's view of time as articulating the non-finite movements of *Geist*. Thus, while there might seem to be an opposition, which Derrida calls in question, between the vulgar and the original, between the derived and the more basic notions of time, there is also an opposition between a temporality, thought in terms of an analytic of finitude, and a temporality thought in terms of Hegel's onto-theology. Husserl's contribution is to have attempted to think the infinities of time on the basis of the finitude of a living present, through the formalisation of a meaning intending, the fulfilment of which can be understood to be indefinitely postponed. Heidegger, however, breaks off his analysis, suggesting to Derrida, as he remarks in his paper '*Ousia* and *Gramme*: Note on a Note in *Being and Time*', that there perhaps is only the vulgar concept of time. Thus any more ambitious thinking of time will have to go by some route other than that of conceptualisation. Before wagering this hypothesis, it is worth exploring the extent to which Husserl and Hegel, as read by Derrida, already move the thinking of time out of the orbit of conceptuality, in the direction of a thinking of generativity, emergent as a relation between speed and limit.

Derrida's reading of Hegel goes against the grain of a Hegel of completion. This is evident in the manoeuvre in *Glas: What Remains of Absolute Knowledge* (1974), of reading Jean Genet's figures and fictions to disrupt the subordination of the family within the state, and of religion within history. It is expedited by his reception of Hegel, through de Man, for whom Hegel's *Aesthetics* is to be read independently of its subordination in Hegel's system to the moment of religious mediation. Thus, for de Man, the idealisations of aesthetics are prone to a reversal into an affirmation of materiality, to which Hegel's text strictly is not subject, if held in place by its own invocation of the truth of religious transcendence. There is thus a line of tension between Derrida's affirmation of de Man's deployment of a Hegel, without religion, and Derrida's contrary line of reading Levinas, to affirm a religion resistant to any Hegelian colonisation. The disjunction, theology without religion, religion without theology has the effect of underlining a fracturing of the Hegelian system, and, for de Man at least, the place of theology in making sense of the possibility of sense, is passed to literary criticism, or an aesthetics, thus released from onto-theology.[33] In 'Typewriter Ribbon', Derrida returns to the engagement with de Man, to set out a different context in which to approach the question of time, marked by a reception of de Man, Lacan and Freud: 'this was an event, perhaps an interminable event' (D: 2001, WA p. 71). De Man invokes a time, which is not temporal, in the conception of history, apparently circumventing the difficulties both of the Nazi inflection in Heidegger's notion of historicality, and the equally problematic Christian specificity of the notions of conversion and confession.[34]

It is a move which Derrida traces out as parallel to Rousseau's de-Christianising of the gesture of confession, of which he writes in 'Typewriter Ribbon':

> Commitment to the future, toward the future, promise, sworn faith (a risk of perjury, promising never to commit perjury), all these gestures present

> themselves as exemplary. The signatory wants to be, he declares himself to be at once singular, unique *and* exemplary, in a manner analogous to what Augustine did in a more explicitly Christian gesture. Rousseau also addresses God, he invokes God and, like Augustine, he uses the familiar '*tu*' form of address.
>
> (D: 2001, WA p. 140)

Derrida moves by way of a discussion of the 'last will' of a last will and testament, a last word, and the temporal structure of pardoning:

> *I forgive you* has the structure of the last word, hence its apocalyptic and millenarian aura; hence the sign it makes in the direction of the end of time and the end of history. We will later get around to this concept of history, which de Man wants to link no longer to time ('History is therefore not a temporal notion', as he will say in 'Kant and Schiller'; it has nothing to do with temporality) but to 'power', to the 'event', and to the 'occurrence'. It corresponds to 'the emergence of a language of power out of a language of cognition').
>
> (D: 2001, WA p. 100)

In question here is the relation of this emergence: whether the language of cognition can remain neutral once this emergent language of power has made itself evident, all the more germane granted the need to respond to the posthumous emergence of de Man's proto-fascist writings. The notion of 'event' here is one of a break in time, detached from its origins in making sense of religious experience. To make sense of this it helps to turn again to Benjamin's separation of time from history, and to consider the role of a reading of Hölderlin, in assisting Heidegger to separate his thinking from its debt to his Christian orientation.

5.2 Time in the name of the other

> The Kantian postulates do not postulate a time after time. Rational hope is not for those who, in time, await events that would come to fill hope's void. For Husserl, intuition fills a signitive aim, as though, one day, what is hoped for would have to make itself known. For Kant, that is impossible: time is the form of sensibility and belongs to the understanding's constitution of phenomenal objectivity. If rational hope were to be fulfilled, if it had to make itself known at a certain moment, that would mean that immortality would have a temporal fulfilment, known in the manner of the phenomenon. But such a contact with the absolute is excluded by the *Critique of Pure Reason*. The rational hope is a hope that cannot be compared with hope in time.
>
> (L: GDT, p. 66)[1]

In his 1998 paper, 'Typewriter Ribbon: Limited Ink (2)', Derrida introduces de Man's notion of the atemporality of history by marking a difference between a time and a history set out in terms of progression and regression, assuming a constituted continuity, and a time and a history, to be set out in terms of break points, and of what has come to be called 'event':

> We must also keep in view a certain concept of *history*, of the historicity of history, so as to trace its intersection with this logic of the textual event as material inscription. When it is a matter of this structure of the text, the concept of historicity will no longer be regulated by the scheme of progression or of regression, thus by a scheme of teleological process, but rather by that of the event, or occurrence, thus by the singularity of the 'one time only'. The value of occurrence links historicity not to time as is usually thought, nor to the temporal process but, according to de Man, to power, to the language *of* power and to language *as* power. Hence the necessity of taking into account performativity, which defines precisely the power of language and power *as* language, the excess of the language of power or of the power of language over constative or cognitive language.
>
> (D: 2001, WA p. 118)[2]

Derrida adds, after a further citation of de Man's text:

> This hyperbolic provocation, in the style of de Man, certainly does not negate all temporality of history. It merely recalls that time, temporal unfolding is not the essential predicate of the concept of history: time is not enough to make history.
>
> (D: 2001, WA p. 118)

Now what is odd about this gloss is that it makes a weaker claim than de Man, or Freud, or indeed Husserl, are inclined to make, and in doing so, it draws on a distinction between essential and inessential predicates. The claim must rather be that there is a hyperbolic time, another time, as marked up by Derrida in an essay 'Tense', paralleling the 'other beginning', invoked by Heidegger in the later text, *Of the Event: Contributions to Philosophy*.[3] This 'other time' must however take a form other than that of a concept, granted Derrida's conclusions concerning the unavailability of any concept of time, other than the so-called 'vulgar', ordinary, or naturalised concept of time, as discussed by Heidegger in *Being and Time*.

Derrida continues in the next paragraph to give the following gloss on a contrast between an 'eventness of an event' and a mediating dialectical process, a comparison which seems implicitly to play Heidegger off against Hegel. This becomes more challenging once it is read in the light of the citation at the start of this section, from Emmanuel Levinas, where the contrast may be thought of as one between the multiple teleologies, set in play by Husserlian primordial impressions, and a unified Kantian teleology, the contrasting principles of which might be thought to underpin the differences between Heidegger and Hegel.

> De Man distinguishes the eventness of events from a dialectical process or from any continuum accessible to a process of knowledge, such as the Hegelian dialectic. No doubt he would have said the same thing of the Marxist dialectic, I presume, if the heritage and thought of Marx could be reduced to that of the dialectic. He also specifies that the performative (the language of power beyond the language of knowledge) is not the negation of the tropological but remains separate from the tropological by a *discontinuity* that tolerates no mediation and no temporal scheme.
>
> (D: 2001, WA p. 118)

The tropological moment of analysis marks an activity of mapping systems of figuration or tropes, which permit continuities of thought cognition and language use to be sustained, across shifts of register and scale. This is to be read in parallel to Freud's topographies of consciousness, which intimate irrecuperable discontinuities between instances of consciousness, while permitting transitions across those discontinuities. In place of a temporal scheme, there emerges a notion of temporality, rethought as a modality of faltering rhythms, syncopated by interruption and *caesura*, where transitions across division are secured by artifice.

Derrida introduces a difference between the excuse, which can be recuperated into a temporal schema, at least as performed by Rousseau's text, and the event of forgiveness, which cannot thus be recuperated.[4] The gap in which such a difference may arrive is thought of as the *caesura*, emphatically introduced by Hölderlin as a marker in his writing and translations, for the arrival and departure of the gods, now to be rethought without this commitment to a thinking of divinity. The painful context of this, in which de Man, from an afterlife, asks his Jewish friend for forgiveness may be remarked:

> It remains the case that the performative, however foreign and excessive it may be in relation to the cognitive, can always be re-inscribed, 'recuperated' is de Man's word, in a cognitive system. This discontinuity, this event as discontinuity, is important for us if only because it will allow us to go beyond the excuse and come closer to the event of forgiveness, which always supposes irreversible interruption, revolutionary caesura, or even the end of history, at least of history as teleological process.
>
> (D: 2001, WA p. 118)

This invocation of a revolutionary *caesura* connects back to the work of Lacoue-Labarthe, on tragedy and the genesis of speculative thought, in the essay 'The Caesura of the Speculative'.[5] There he writes:

> The question I am posing therefore has to do with the possibility in general of a demarcation of the speculative: of the general logic of differentiation, of the ordered contradiction, of the exchange or the passage into the opposite as the production of the Same, of the *Aufhebung* and of (ap)propriation, etc.
>
> (p. 211)

This passage into a production of sameness is to be contrasted to the Derridean 'absolute as passage', where no such sameness can return. The logic of sameness is put in question in the writings of Hölderlin on tragedy, which disrupt the closures of German idealism, in the work of Hölderlin's contemporaries, Schelling and Hegel. Of this Derrida writes in his introductory remarks to the English translation of Lacoue-Labarthe's text, under the title 'Desistances':

> When in Sophoclean tragedy, it marks the withdrawal of the divine and turning back of man toward the earth, the *caesura*, gap or hiatus plays at and undoes mourning. A *Trauerspiel* plays at mourning. It doubles the work of mourning; the speculative, dialectic, opposition, identification, nostalgic interiorization, even the double bind of imitation. But it doesn't avoid it.
>
> Gap or hiatus: the open mouth. To give and receive. The *caesura* sometimes takes your breath away. When luck is with it, it's to let you speak.
>
> (p. 42)

Thus the material conditions of speech, or interrupting breathing, are invoked to mark a physiological clue for thinking the interruption of processes required for the arrival of the future.

Earlier in the essay, Derrida remarks a connection between this hiatus and the double law, which has come to be known as deconstruction, a law here glossed as marked out in the hiatus between the law of the three plus or minus one, and the law of the symptom.

> If I posited or proposed anything at all in this writing, it would not be merely a theory, or even a practice of the double bind according to the measure of an immense tradition – a tradition marked in its rhythm, continued and interrupted by what all these proper nouns appear to sign. I present myself, or rather write myself, sign my own distance, the impossible itself, as an experience of the double bind, the poetic experience of the double bind. Double constraint, double law, knot and caesura of the double law, the law of the double. The knot and the caesura, the obligation and the break – that's rhythm. And is rhythm not the double law – and vice versa? The task would be to think this. And the supplementary ring of this deconstruction would be nothing other, would have no other modality than this double constraint that no dialectic could ever overcome.
>
> (p. 34)

The *caesura* of desistance then marks up how a positing of existence runs up against a limit, subverting its own programme, by revealing that limit; the attempt to delimit a meaning objectively brings into play the meaning of the existence of the one delimiting that meaning, thus bringing into play an oscillation between attention to objective meaning and attention to its conditions, in the existence of the one who assigns meaning.

The relation between excuse and forgiveness, one as recuperable within cognitive continuity, and the other as a limit or turning point of temporal organisation, forms a schema through which Derrida poses a question to the activity of writing, and to the activity of constituting bodies of work through processes of reception, and indeed inheritance. He introduces a notion of the survival of the work, as a mode of living on, less than that of full temporality, a mode which, by implication, Heidegger would call 'poor in world':

> We are seeking in this way to advance our research on the subject of that which in forgiveness, excuse or perjury *comes to pass, is done, comes about, happens, arrives* and thus that which, as event, requires not only an operation, an act, a performance, a *praxis*, but an *oeuvre*, that is at the same time the result and the trace left by a supposed operation, an *oeuvre* that survives its supposed operation and its supposed operator. Surviving it, being destined to this sur-vival, to this excess over present life, the *oeuvre* as trace implies from the outset the structure of sur-vival, that is what *cuts* the *oeuvre* off from the operation. The *cut* assures it a sort of archival independence or

> autonomy that is quasi-machine like (not machine-like but *quasi*-machine like) a power of repetition, repeatability, iterability, seriality and prosthetic substitution of self for self.
>
> (D: 2001, WA p. 133)

This survival, affirming a quasi-life beyond life, is a correlate of the trace-structure of writing, as articulation of meaning. It is the *survivre*, the displaced and deferred intensified life of poetry, analysed in 'Living On – Border Lines'. In the interview *Apprendre à vivre enfin: entretien avec Jean Birnbaum* (2005), Derrida states that this structure of the *survivre* is articulated through autobiography, and held in place by the twin movements of the trace and the testament.[6]

The analysis is brought to the point of formulating the following questions:

> To say in this way that the *oeuvre* institutes and constitutes an event is to register in a confused way an ambiguous thing. An *oeuvre* is an event to be sure; there is no *oeuvre* without singular event, without textual event if one can agree to enlarge this notion beyond its verbal or discursive limits. But is the *oeuvre* the trace of an event, the name of the trace of the event that will have instituted it as *oeuvre*? Or is it the institution of this event itself?

Derrida responds to these questions thus:

> I would be tempted to respond, and not only so as to avoid the question: both at once. Every surviving *oeuvre* keeps the trace of this ambiguity. It keeps the memory of the present that instituted it, but, in this present, there was *already* if not the project at least the essential possibility of this cut – of this cut in view of leaving a trace, of this cut whose purpose is survival of this cut that sometimes assures survival even if there is not the purpose of survival. This cut is at once a wounding, an opening, and the chance of a respiration, and it was in some way already there at work, *à l'oeuvre*. It marked, like a scar, the living originary presence of this institution – as if the machine, the *quasi*-machine were already operating, even before being produced in the world, if I can put it that way, in the vivid experience of the living present.
>
> (D: 2001, WA p. 133)

There is here a juxtaposition between the writing of a living autobiographical animal, and its continuation, surviving as a quasi-machine, as a vehicle of meaning, after the death of its writer. The quasi-machine producing meaning is in place in advance of the arrival of meaning in a living present, and the living present is shown to be dependent on the timelessness of the machine. The organic becomes inorganic, and living meanings are materially inscribed, deploying a set of artificial relations, amongst others, the figurations analysed in de Man's accounts of figure and tropes.

These relations between the organic and the inorganic, the machine and the event of innovation suggest a need to return to the Kant of the Third Critique.

However, here, Derrida's responses to de Man and to Rousseau stand in for such a return, which is thus complicated by the necessary detour, unfixing Kant's certainties. Derrida does, however, invoke the Kantian figure of antinomy, in the opening passages of the essay:

> it would be necessary in the future, (but there will be no future except on this condition) to think *both* the event and the machine as two compatible or even indissociable concepts. Today they appear to us to be antinomic. Antinomic because what happens ought to keep, so we think, some non-programmable and therefore incalculable singularity. An event worthy of the name ought not, so we think, to give in or be reduced to repetition.
>
> (D: 2001, WA p. 72)

This appearance of antinomy conceals the more basic form of the aporetic structure of time, for this, unlike the antimonies of Kant, is an antinomy without resolution. The second part of the *Critique of Judgment* (1790) sets out an antinomy concerning how to think nature, as either a purposive structure or as mechanism. Kant's solution to the antinomy, breaking the distinction between the mechanism of nature and the purposiveness of intentions, is to hypothesise an orderliness of nature, as an idea towards which nature tends, and in accordance with which it is conceived, but which cannot be used as constitutive of natural phenomena.[7] Derrida poses a different version of the antinomy, by identifying a tension between affirming an initiating writing activity as the source of meaning, and a continuing generation of meaning, long after the resulting writings have ceased to have meaning for their writers. Derrida restates the problem, in relation to making sense of the notion of an inaugurating event:

> To respond to its name, the event ought above all to *happen* to someone, to a living being who is thus *affected* by it, consciously or unconsciously. No event without *experience* (and this is basically what experience means), without experience, conscious or unconscious, human or not, of what happens to the living.
>
> It is difficult, however, to conceive of a living being *to* whom or *through* whom something happens without an affection getting *inscribed* in a sensible, aesthetic manner right on some body or some organic matter.
>
> (D: 2001, WA p. 72)

Affects are inseparable from the organic matter in which their effects are inscribed. He continues: 'Why *organic*? Because there is no thinking of the event, it seems, without some sensitivity, without an *aesthetic* effect and some presumption of living organicity' (D: 2001, WA p. 72).

The address to de Man and to Rousseau, in place of an expected reading of Kant, is further disrupted by the thought that, as much as Kant's account of self-affection and of transcendental aesthetics, also relevant here is Husserl on affect and aesthetics, for he, like Kant, has an account of a transcendental aesthetic as the set of conditions of possibility for the inscription of affects. These substitutions and displacements are then to be further interrupted by adducing the following startling remarks from a text appended to Husserl's *On the Phenomenology of Internal Time Consciousness*: 'But the question is, whether it makes sense, in the strong and genuine sense, that the constituting appearances belonging to time consciousness (to the consciousness of internal time) themselves fall into (immanent) time' (text 54, HUE 4, p. 380).[8] The penultimate paragraph of this same text makes the parallel point with respect to consciousness:

> We find moreover that when we do pay attention to something, something is always already 'appearing' – the style of attention always runs through and across an intentionality. But if I direct my regard towards an actual momentary phase of the flow? But we should seriously consider whether we must assume such an ultimate consciousness, which would necessarily be an 'unconscious' consciousness; that is to say, as ultimate intentionality, it cannot be an object of attention (if paying attention always presupposes intentionality already given in advance), and therefore it can never become conscious in this particular sense.
>
> (HUE 4, p. 394)

This Husserl, then, combines de Man's question to a time, which may not itself be temporal, to Freud's question to a form of consciousness, which may itself not be accessible to consciousness. A gap thus opens up between the Husserl of the self-evident *apriori* structures of time, and the Husserl of these deeply paradoxical considerations, an aporetic time thought by an unconscious consciousness of non-simultaneity. The possibility of pursuing Derrida's thinking of the writing machine as both organic and inorganic, as a successor to and disruption of Kant's distinction between mechanical and final causes, must go by way of Derrida's tracing and concealment of the hypotheses of Freud and of de Man, in Husserl's phenomenology, indicating a concealed and perhaps irretrievable origin for time and for consciousness.

The temporality analysed by de Man structures the articulation of materially given texts and objects of aesthetic attention in a public domain of critique. In the essay 'The Rhetoric of Temporality' he identifies the turns, decelerations of tempo and velocities of figuration in relation to discursive flows, and shows temporality as also in process of being constituted in the deployment of language, which is itself articulated through and as modes of temporalisation.[9] The analyst, here, however, remains offstage like a puppet master, manipulating texts, artefacts and meanings like marionettes, into an appearance of life. Husserl's analyses by contrast seek to reveal that, without a living process, of the bearer of the intelligence which performs the tasks of analysis, there is no meaning and nothing to analyse. He places the focus of attention on the processes of layering, making up such performances and setting out the connections between such bearers of intelligence, in the analyses of inter-subjectivity. As a result, Husserl can reveal a labyrinthine series of relays between levels and instances of consciousness. Husserl's analyses reveal non-intentional levels of synthesis, in passive synthesis, underlying the development of the active syntheses of deliberate intent. There thus arises the paradoxical notion of a passive, or non-conscious intentionality and indeed an intentionality without intent, in parallel to Kant's purposiveness without purpose, but with even greater disruptive effects, on the supposed unity of concepts of self and consciousness, and on the singleness, linearity and orderliness of time.[10]

In the second, shorter essay in *Rogues: Two Essays on Reason*, 'The "World" of the Enlightenment to Come (Exception, Calculation, Sovereignty)', Derrida analyses the role of re-reading the texts of Kant and of Husserl, for the task of becoming the 'responsible guardians we must be of this heritage'.[11] He remarks:

> We must ask ourselves whether in their very historicity (for there is an undemonstrable thought of history in Kant and in Husserl, and even a place for a certain history of reason) these great transcendental and teleological rationalisms generate a thought of – or expose themselves to – that which comes.
>
> (D: 2003, RS p. 135)

Thus Derrida, in 2003, supposes that the teleologies of Husserl and of Kant are open to a certain subversion, and redirection, as indeed his readings of Heidegger and Freud have tended to show. As the invocation of Freud disrupts the reading of Husserl, so the readings of Freud and of Heidegger, especially but not only in *The Post Card*, underline a *restance*, as an irrecuperable layer within the structure of intentionality, such that there cannot be a complete recuperation of the conditions for thought, in what is then made explicit. This underlines the differences between the requirement of perspicuity in Hegel's analyses and the irreducible function for Husserl's phenomenology of pre-predicative impressions. Similarly, Heidegger's analysis of Dasein reveals irrecuperable origins for the trajectory of a history of being, which no amount of Hegelian transposition can bring back. The role of these critiques of teleology and of transcendentalism, those of Heidegger and Freud, and indeed of Derrida, is both to display the transcendental and teleological commitments of the rationalisms proposed by Husserl, by Hegel, and by Kant, and to pass them on, while also putting them into question as an inheritance.[12]

This interrogation is complicated by the suspicion, developed in the series of texts delivered as lectures at the conference '*L'animal autobiographique: autour de Jacques Derrida*', at Cérisy-la-Salle in 1997, of the tendency in the philosophical tradition to think the human as close to the divine, and to secure that connection by seeking to detach the human from animality.[13] The argument of 'L'animal autobiographique' is articulated through the semantic variations animal, *animot*, invoking words and names, and *animal*, as including the notion of evil, *le mal*. The human is the animal through whom words and names, and evil, God's other, arrive in the world. The longer text, published only in 2006, reveals how Derrida sets up two series of writings, to be read in parallel. There is the series: Descartes, Kant, Bentham, Heidegger, Levinas, Lacan, to which he proposes to attend, in the interests of interrogating the tendency in this philosophical tradition, to sacrifice the animal, in the human, in order to elevate the capacities of the human as rational. He also reads the poetry of Hölderlin and Celan, of Valery and of Rilke, to interrupt the intimations drawn from the first series, and to re-connect to the thinking of the animal to be found in Nietzsche's texts, and in Kafka's transformations. These enquiries are developed in accordance with an emergent logic of the symptom, and of a law of the three, plus or minus one, with which Derrida already in *Glas: What Remains of Absolute Knowledge* (1974) sets about disrupting the necessity of the Hegelian synthesis and resolution (*Aufhebung*) of opposing theses. The logic of the symptom permits a reversal of the temporal order, between supposed effect and supposed cause, since the arrival of the affect, long after the event which precipitates it, compels a return to the occasion of that event of its concealed inscription, to reveal, in a work of phenomenology, that concealed impact. In this way a logic of responses and questions can be set up as counter to the more customary logic of premises, antecedents and consequents. The response, or the affect reveals that when the question has been posed and become formulable, the system of registration is revealed to have undergone a transformation, such as to permit a tracing out of the impact of that delayed event. Thus the supposed continuity of time is broken by shifts brought about by alterations in the structures of registration, themselves.

The domain of well ordered, quantifiable entities is opened out on to the migrations of animality, and the instabilities of an oneiric logic of appearances, temporarily stabilised such as to reveal an image, as object of awareness, but not of the order of the medium-sized, dry goods, maintained in, and maintaining an independently constituted stable, spatio-temporal realm. Where for Husserl, the logic of what appears is to be disambiguated by careful attention to the evidences of givenness, and in accordance with a logic, of wholes and parts, of hierarchies and regionalisation, this combination of logics as preferred by Derrida, provides only temporary ontological stabilisation, dependent as much on habit and performative emphatics, as on any matter of fact concerning what there is, or can be thought to be. The work of repetition in producing a stable ontology and a continuous space-time is disrupted by the event of iteration, in which the system of registration alters, disjoining spatio-temporal continuity. Thus iteration, on the basis of a changing order of things, is to be contrasted to repetition presuming a stable ordering. This then is an ontology of impermanence, and it is reliant on the work of the legitimating fictions, explored by Derrida in terms of Michel de Montaigne's account of a 'mystical authority', not just of the law, as social-political institution, but of ontology more generally.[14] Husserl's notion of intentionality is thus transposed from the stabilising framing of a relation between an intending consciousness and its objects of awareness, into a meaning-making relation, suspended in the texts, which subsist independently of their composition and of their legibility, but which are meaningful only when reactivated, through reading, at which point meaning is re-assigned to them. Derrida thus transposes the notion of intentionality into that of a textuality, as autobiography, where the animal writes itself into determinacy, but only as following on from the other, through a line of selective variation, and inflected through a notion of an intermittent transmission, in accordance with the modalities of a 'perhaps' and a 'maybe'.[15]

The paradoxes of a time, which is no longer to be thought as chronological time, are more perspicuous once a distinction between a time, or temporality, as inner time consciousness, and time as an articulation of Augustine's distension, is in place. For Husserl's inner time consciousness is not necessarily subordinated to a divine time of eternity, whereas this Augustinian distension must be understood as an imperfect or inauthentic version of a divine time of eternity. Kant's account of time can then be understood as an unhappy compromise between the two, not willing to help itself to Augustinian certainties concerning a divine revelation, but also unwilling to affirm a Husserlian asceticism. The Augustinian tradition is reaffirmed and brought into the context of analyses of history and meaning in the complex movements traced out in Paul Ricoeur's *Time and Narrative.*[16] Husserl's account of time by contrast has more in common with the disruptive time of the Heideggerian *Ereignis*, of a withdrawal, which presents an appropriable order. This transposes Biblical accounts of a divine arrival in human time, as disruption, into an account of the paradoxical nature of an ontological transformation, of which only fragmentary traces make themselves present. Both Husserl's inner time consciousness and Heideggerian historicality are implicit reference points for both de Man, and for Derrida, in his reading of him.

For both Husserl and Heidegger, the relation between Augustine and Aristotle on an inauguration of a questioning, and an understanding of time, as distinctively human achievements, are critical determinants of the parameters of all subsequent enquiry. However, there is no presumption that there is an orderly fit between a Greek inheritance, a Christian gloss on a relation between divine and human time, and history, as falling into a pre-established temporal continua.[17] Where for the Aristotle of the *Physics*, time and movement are continua, and for Augustine and Hegel, these continua are to be grounded in a time made complete by divine intending, for Husserl and for Heidegger these relations are left open. Ricoeur seeks to close the gap again, by articulating a notion of narrative continuity, whereas, with Derrida, one of the principal effects of his writing and his interventions in the transmission of the tradition is to underline that this mythological set of fits can be maintained only at the price of welding Judaism together with Christianity, and by eliding the difference which is Islam altogether. The readings of Levinas and of Marion, on one side, and of Nietzsche and Freud on the other, produce a spectrum of ways of setting out the relations between the time of philosophy and the time of religious illumination, which are thus variously linked up and disjoined. For the more obvious parallel between history, as a time which is not in time, is not with a time of divine intending, before creation, but with the Freudian notion of the unconscious, as articulating a time which is not in time, at least if time is presumed to be an orderly continuum, commensurable with clock time and times of measurement.

Freud insists in 'The Unconscious' (1915), that unconscious processes are not temporal, if temporality is understood to be the time of continuous conscious experience, articulated into a sense of personal identity:

> The processes of the system Ucs. are timeless, i.e. they are not ordered temporally, are not altered by the passage of time, they have no reference to time at all. Reference to time is bound up once again with the work of the system Cs.
>
> (PFL, vol. 11 p. 191)[18]

Freud returns to this feature of his topographies on a number of occasions and, in 'Beyond the Pleasure Principle' (1920) explicitly makes a connection to an implicit challenge to the Kantian conception of time:

> At this point I shall venture to touch for a moment upon a subject, which would merit the most exhaustive treatment. As a result of certain psycho-analytic discoveries, we are today in a position to embark on a discussion of the Kantian theorem that time and space are 'necessary forms of thought'. We have learnt that unconscious mental processes are in themselves 'timeless'. This means in the first place that they are not ordered temporally, that time does not change them in any way and that the idea of time cannot be applied to them. These are negative characteristics, which can only be clearly understood if a comparison is made with *conscious* mental processes.

He then contrasts this to the dominant notion of time, which draws exclusively on the relation 'Pcs. Cs', the continuum of the preconscious, and the conscious, to the exclusion of whatever non-serial, non-naturalised temporality operates in the system 'Ucs.', the unconscious:

> On the other hand, our abstract idea of time seems to be wholly derived from the method of working of the system Pcpt. Cs. and to correspond to a perception on its own part of the method of working. This mode of functioning may perhaps constitute another way of providing a shield against stimuli.
>
> (PFL, vol. 11 pp. 299–300)

This remark could well be made the focus for Derrida's readings of Freud's 'Beyond the Pleasure Principle', in 'To speculate – on "Freud"' (1978), connecting it back to the diagnosis of refractory temporal effects in 'Freud and the Scene of Writing' (1966).

In that earlier text, Derrida brings a section on 'The print (or impression) and the original supplement' to a close with the following observation: 'We ought perhaps to read Freud the way Heidegger read Kant; like the *cogito*, the unconscious no doubt is timeless only from the standpoint of a certain vulgar conception of time' (D: 1978, WD p. 214). The print or impression, *Eindruck*, is thus in some sense prior to the *Ausdruck*, the expression, in which that impression then comes to expression. This introduces a temporal impression before the express articulation of temporality. This structure of disruptive anticipation is explored at some length in *Archive Fever: A Freudian Impression* (1994) in which the possibility of an expression retroactively bringing a prior impression into existence is pursued, through an elaboration of the notion of the archive and of the logic of the three plus or minus one. This disrupts the fixity of the logic of before and after, key to certain conceptions of time.[19] In the earlier essay, Derrida writes of the Freudian impression:

> The conscious text is thus not a transcription because there is no text *present elsewhere* as an unconscious one to be transposed or transported. For the value of presence can also dangerously affect the concept of the unconscious. There is no unconscious truth to be rediscovered by virtue of having been written elsewhere. There is no text written and present elsewhere which would then be subjected, without being changed in the process, to an operation and a temporalization (the latter belonging to consciousness, if we follow Freud literally) which would be external to it, floating on the surface. There is no present text in general, and there is not even a past present text, a text which is past as having been present. The text is not conceivable in an originary or modified form of presence.
>
> (D: 1967, WD p. 211)

This original impression provides an alternative to the account of originary impressions which form the basis of the Husserlian account of meaning, and in so far as Husserl can be shown to move from an account of impression, as given in an originary presence, towards such a textual view, Derrida can be supposed to affirm a Husserlian lineage, as opposed to constantly working to disrupt the Husserlian account. Derrida continues: 'Everything begins with reproduction. Always already: repositories of a meaning which was never present, whose signified presence is always reconstituted by deferral, *nachträglich*, belatedly, *supplementarily*: for the *nachträglich* also means supplementary' (D: 1978, WD p. 211). To this series of terms, Derrida will subsequently add the notion of the prosthesis, and of language as the originary prosthesis. In a footnote to this text, Derrida adds:

> The concepts of originary *différance* and of delay (*retard*) are unthinkable within the authority of the logic of identity or even within the concept of time. The very absurdity betrayed *by the terms* provides the possibility, – if organized in a certain manner of thinking beyond that logic and that conceptuality. The word 'delay' must be taken to mean something other than relation between two 'presents'; and the following model must be avoided: what was to happen (should have happened) in a (prior) present A, occurs only in a present B. The concept of originary *différance* and originary 'delay' were imposed on us by a reading of Husserl.
>
> (D: 1967, WD p. 329)

thus firmly linking the reading of Freud to an elaboration of what was found lacking or unstable in Husserl.

Cutting across Derrida's writings then there is an ongoing engagement with Freud's questioning of time and meaning as not adequately figurable within the parameters of a continuous series. This longitudinal preoccupation with time erupts on to the surface of the other readings of time, from time to time, and poses the question whether it is more suitable to read Derrida's texts as grouped together by temporal proximity, as with the texts on de Man and on animality, which were initially presented in 1998 and 1997, respectively, or whether such a longitudinal section is the more instructive approach. Such a reading could follow the line of articulation from the essay 'Freud and the Scene of Writing' from 1966, down to the 'States of the Soul of Psychoanalysis' delivered on his birthday, 16 July 2000. However, this would necessarily go by way of the as yet incompletely published seminars on Freud from 1975, entitled '*La vie la mort*', life death, part of which appears in *The Post Card*, with the following explanatory note:

> The text on whose borders this discourse would be attempting to maintain itself is Freud's 'Beyond the Pleasure Principle' [in volume 18 of the Standard Edition; all references in the text by page number]. In effect I am extracting this material from a seminar which followed the itinerary of the three rings. Proceeding each time from an explication with a given text of Nietzsche's, the seminar was first concerned with a 'modern' problematic of biology, genetics, epistemology, or the history of the life sciences (readings of Jacob, Canguilhem etc.). Second ring: return to Nietzsche, and then an explication with the Heideggerian reading of Nietzsche. Then here, the third and last ring.
> (D: 1980, PC p. 259)

This reading must be postponed, and instead of it, I shall pursue some of the repressed lines of the deflected encounter with Kant, as indicated by the citation at the head of this section, and thus returning to the relation between Husserl and Kant indicated in the opening pages of these enquiries. It will thus be possible to arrive at a provisional account of the transformation of Kant's account of time, consequent on the development of phenomenology by Husserl, and the responses to it, of Heidegger, Levinas, Marion, and Derrida.

Levinas' lectures *God, Death, Time* were delivered in 1975, when Derrida was considering his views on Freud, in relation to a reading of Nietzsche, and on Heidegger's reading of Nietzsche. Just as Levinas proposes a reading of Kant, which is not to be subordinated to Heidegger's question of being, so Derrida considers a reading of Nietzsche, which is not to be subordinated to this questioning either. At this time, furthermore, Levinas wrote the short paper in which the little dog at the prison camp gate, to whom the prisoners gave the name Bobby, is described as the last Kantian in Nazi Germany.[20] These lectures, given at the Sorbonne, begin thus: 'In question in this course is, above all, time – this is a course on the duration of time. The word duration is chosen for several reasons' (L: GDT, p. 7). He then adduces four:

- It indicates that we will not be posing here the question 'What is ... ?'. In the same way, in an unpublished lecture prior to *Being and Time*, Heidegger said that one cannot pose the question 'What is time?' because one then immediately posits time as being.
- There is no action in the passivity of time, which is patience itself (this being said as against the intentional approach).
- The word avoids ideas of flux and of flowing, which make us think of a liquid substance and announces the possibility of a measure of time (time measured, or clock time, is not the authentic time). As temporalization – *Zeitigung* – the word duration avoids all these misunderstandings and avoids the confusion between what flows within time and time itself.
- It is a term that above all would leave to time its own mode.

(L: GDT, p. 7)

The notion of duration connects to the analyses of time offered by Bergson, and, by implication, Bachelard's reception of Bergson's notion of duration, in *Dialectic of Duration* (1950), which Levinas introduces in lecture eleven.[21] While Levinas explicitly disavows a thinking of time in terms of intentionality, there is an acceptance from Husserl and Heidegger of the notion of a *Zeitigung der Zeit*, a temporising of time, which shifts the thinking of time from a deliberate thematisation of an essence, *die Zeitlichkeit*, to focusing on an auto-hetero-genesis for time, *die Zeitigung*, in the manner of a passive synthesis. For time, *die Zeit*, is not identical to its genesis, as *die Zeitigung*, but it is not wholly distinct from it either. The relation between the two, between time and its genesis, is to be thought in terms of a relation of temporalising, as a process neither of deliberate intending, nor of mechanical, causal or natural inevitability. It is to a refinement of this hybrid notion of a genesis of time that I understand Derrida to be making a major contribution.

In the second lecture, given on 14 November 1975, Levinas invokes an anteriority 'more ancient than any *a priori*', an 'immemorial diachrony', that cannot be brought back into experience. This anterior time is linked to a relation with death, obtaining prior to any experience:

- The relation with death, more ancient than any experience, is not the vision of being and nothingness.

Intentionality is not the secret of the human. The human *esse*, or existing, is not a *conatus*, but disinterestedness and *adieu*.

Death: a mortality as demanded by the duration of time.

(L: GDT, p. 15)

Thus Levinas distances himself from Sartre, and indeed Hegel, as implied in the invocation of being and nothingness, and from any re-inscription of intentionality into the rationalist metaphysics of Spinoza, implied in the invocation of the *conatus*. Disinterestedness and the *adieu* separate off the human *esse* from theories locating it solely in terms of being, *inter esse*. It flags up instead a directedness from and to God, in the '*adieu*', as a greeting, a leave taking and the bidding of farewell, as the commendation to one's God, and a reception of a command from one's God. The fourth lecture, entitled: 'An Obligatory Passage: Heidegger', indicates Levinas' continuing reservations with respect to affirming a debt to Heidegger.[22] Not, however, until lectures twelve and thirteen does Levinas mobilise a non-Heideggerian reading of Kant against that of Heidegger, under the title 'The Radical Question: Kant against Heidegger', and it is the fourteenth lecture which begins with the citation given at the beginning of this section of my enquiries, concerning Kant's postulates of empirical thought.

In the twelfth lecture, Levinas identifies a contrast between his notion of duration, and the distinction between authentic or originary time, and that between a time of the fall, and an everyday drifting, which emerge from his characterisation of Heidegger's distinctive mode of enquiry in *Being and Time*:

> From this came a characterization of time, that is, of the originary time of which our everyday time is but a fall or – to use a word that does not confer value on this – a drifting. To take death on in anxiety as the possible is what is originally to-come, or future (*a-venir*) (which is itself to-be); a future engaged in a past because anxiety is an affectivity containing the always already, and the always already-there. The entire structure of time is drawn from the relationship to death, which is a modality of being.
>
> (L: GDT, pp. 57–58)

Derrida is thus with Levinas in a critique of reducing the *a-venir* to a modality of a 'to-be'. Levinas explores the reading of Kant made by Heidegger in order to disrupt this account of time and death:

> Kantian philosophy was thus reduced by Heidegger, who insisted above all on the *Critique of Pure Reason*, to the first radical exhibition of the finitude of being. But of the four questions posed in philosophy, according to Kant (What may I know? What must I do? What am I entitled to hope? What is man?), the second seems to surpass the first with all the breadth of the last two. The question 'What may I know?' leads to finitude, but 'What must I do?' and 'What am I entitled to hope for?' go further and, in any case, elsewhere than toward finitude. These questions are not reducible to the comprehension of being: they concern the duty and salvation of man.
>
> (L: GDT, pp. 59–60)

Levinas then concludes this lecture with the remark: 'The practical philosophy of Kant shows that the Heideggerian reduction is not obligatory. It shows that there might be, in the history of philosophy, a signification other than that of finitude' (L: GDT, p. 61).

The thirteenth lecture makes a link from Heidegger's account of finite time back to that of Husserlian omni-temporality:

> Everything that could have a meaning, every project, every comprehension, is reduced to this being, and death is the ultimate, the certain, the ownmost and unsurpassable possibility – or again, time goes back (speculatively) to the epic of being as being, which is being-to-death. To this epic can be traced the dimension of time and even that of intemporality or of the ideal (in Husserl, eternal ideality is already an omni-temporality). There is no eternity; eternity is, like linear time, a modality of finite time; it is derived from originary time.
>
> (L: GDT, p. 62)

In contrast to both of these, Levinas invokes a Kantian hope, which is wholly other than that of any motivation arising from a finite existing, destined to end in death. Of this he writes:

> This orientation does not refute being-towards-death, which is, according to Heidegger the presupposition of finitude. There lies the great force of Kant's practical philosophy: the possibility of thinking a beyond of time by way of hope, but obviously not a beyond that would prolong time, to a beyond that *is* (and would be). But neither an everyday derivative of originary time Rather, a rational hope, as though in finite time there opened another dimension of originarity that was not a denial inflected on finite time but something that had meaning other than finite *or* infinite time. The meaning of this hope in despair does not undo the nothingness of death: it is a lending to death of a meaning other than that which it draws from the nothingness of being. It is not to a need to survive that this hope answers.
>
> (L: GDT, pp. 63–64)

Levinas takes this diagnosis as an invitation to perform a turn to God, which is circumscribed neither by the onto-theology, already put in question by Kant, nor by the anticipation of new gods, initiated by Hölderlin and Nietzsche and, in their wake, Heidegger. It is this turn to God which is reprised in Jean-Luc Marion's appropriation of Husserl's phenomenology.

Derrida refuses all three moves, and in his reading of Levinas' *Adieu: To Emmanuel Levinas*, he recasts the religious contraband of the notion of the *adieu* as a dedication, without binding religious commitment. He seeks to make this obligatory passage back through Heidegger to Kant, without the affirmation of the God of Israel, to whom Levinas is again committed by the end of his lectures.[23] For Derrida's writings seek to open an *a-venir* which is to be re-inscribed neither in an onto-theology, nor in an ontology of being, nor in the dedication of an *adieu*, to either old or new gods. The question which remains open is whether Husserl, like Levinas, can hold the future open only by resort to such a dedication, or whether Husserl can be read, in the light of Derrida's challenge, as already holding open a future, without any such immemorial anticipation of a deity. Levinas' future remains open by virtue of a dedication in the name of the father. Perhaps Husserl's time remains open in the name of this other, Derridean

Husserl, or Husserlian Derrida, as a consequence of a certain performative reading. This is not the other as father but the other in me. By returning to Kant, Levinas puts in question Heidegger's decision in favour of an originary finitude of time. Derrida has to approve of this move, to be consistent with the methodological principle of suspending the decision between two competing modes, in this case the thinking of time, as primordially non-finite or as primordially finite, but he has to be opposed to the affirmation of new and old gods. It is indeed for Derrida the virtue of the writings of Freud and of Husserl, as opposed to those of Levinas and Heidegger, that the former, Freud and Husserl, leave open the question of the basic constitution of time, in a way that neither Heidegger nor Levinas are disposed to do.

The Kantian postulates, the Postulates of Empirical Thought, are the fourth set of transcendental principles examined by Kant, in the Transcendental Analytic of the *Critique of Pure Reason*, and they correspond to the categories of possibility, actuality and necessity, and to the forms of judgment: the problematic, the assertoric and the apodictic. The postulates are:

1 That which agrees, in intuition and in concepts, with the formal conditions of experience, is possible.
2 That which is bound up with the material conditions of experience, that is, with sensation, is actual.
3 That which in its connection with the actual is determined in accordance with universal conditions of experience is (that is, exists as) necessary.

(Kant: A 218, B 266)

Here there is a series of determinations following the sequence, formal conditions, material conditions, and universal conditions, which map for Kant on to the categories of possibility, actuality and necessity. For Levinas, the sequences and their mappings are thrown into question by his presumption that in advance of any actuality, there lies the immemorial moment of a divine intending, invocation, or naming, marked by the arrival of the name of the divine, in the manifestation of the face and the subsequent recognition of the claim of the other.

For Levinas, there is an absolute necessity beyond any such notion of necessary conditions of experience. For Derrida, by contrast, there is no such evident givenness of the divine, in advance of a separation of the moments of analysis, but there is the claim from the essay on 'Freud and the Scene of Writing': 'Time is the economy of a system of writing' (D: 1978, WD p. 226). There is an economy of writing which will subsequently be articulated into the critique of the Husserlian time of the thesis, as always already disrupted in advance by the time of delay, in the modes of inscription provide by the prosthetic quasi-machines of writing, whereby the meanings of natural languages are pre-inscribed and kept in suspended animation, even when there is no-one to intend meaning. Derrida continues in 'Freud and the Scene of Writing':

> Writing here is *techne* as the relation between life and death, between present and representation, between the two apparatuses. It opens up the question of technics: of the apparatus in general and of the analogy between the psychical apparatus and the non-psychical apparatus. In this sense writing is the stage of history and the play of the world.
>
> (D: 1978, WD p. 228)

And here history and the 'play' or, perhaps, better 'slippage', of the world take priority over any analysis of time in terms of temporality. This play or slippage of the world inserts a deflection, preventing meaning as intended from arriving at a fulfilment of that meaning, as intended. History is the series of loopings, whereby subsequent registrations of impressions shift the meanings and even transform the inscription systems into something other than they were. Necessity gives way to the happenstance of such deflection and of the chance of its inscription being noted. The absolute as passage installs a priority to a certain chance of deflection, detour and drift. The logic of this drifting is laid out with exemplary clarity in the jointly authored text, *Counterpath: Traveling with Jacques Derrida*, by Catherine Malabou and Jacques Derrida.[24]

These are the effects of the 'perhaps' or 'maybe' which, as a temporising modality of what there is, for Derrida, takes priority over apodictic necessities of various kinds, hypothesised in the tradition. They impose on time, conceived of as smooth continua, or as transparent orderliness, an opacity and swing, the *clinamen*, which constitutes the deviation of any identity from itself. This is articulated by Derrida, in so far as it can be articulated, through an adaptation of the logic of the symptom, which opposes the logic of natural causes as conceived by Kant, and by the logic of the three, plus or minus one, which opposes the Hegelian logic of dialectical sublation and completion. These might be developed to interrupt the working of any law of an excluded middle. They reiterate the move of Husserlian bracketing, the *epoche*, which suspends unexamined ontological commitment, but without the concomitant commitment to the task of arriving at a secure basis on which to reinstall the ordering of world and time, as apodictically given. For Derrida neither time, nor the concept is subject to a dialectical logic of sublation and completion, and what there is, is marked by the fracturing of this nineteenth-century vision of time, philosophy and history, behind which Husserl, and Heidegger, and Derrida seek to reach back, to the less sanguinary, and less smoothly dovetailing notions of time, and of the limit, made available both in the Greek tradition and in the various traditions hidden from view by the various hegemonies of the spirit, enacted by Eurocentric histories of ideas and reflections on philosophy.

Thus where for Levinas the time before the time of human experience is installed in an immemorial diachrony, of a divinely named creation, for Derrida, studying the texts in which this naming as creation is reported does not serve as a prelude for an affirmation of faith. Faith, no more than madness, cannot permit an escape from the system of differences held in place by the finitude which is infinite *différance*. Nor are the texts of a literary displacement of the topics of naming, creation, topology and invention a sufficient source for him for risking an affirmative account of time and history, despite his regard for and repeated readings of de Man. His explorations of the paradoxes of faith, disrupting any presumed autonomy and authority for the writings of reason, do not lead him to affirm anything more than a formally indicated notion of an *adieu*. The enquiry concerning time working through Derrida's texts opens out a series of breaks in the supposed continua of duration and movement, marked up in this part of my enquiries, by a marking up of a time without history and of a history without time; in Part IV, of a religion without theology and of theology without religion; and in Part III, of the scarcely erased traces of a successor relation between the narratives of religious texts and the narratives, denied and affirmed, under the title 'literature'. The manner in which these various lines of enquiry take up and develop the problems to be addressed under the title of the abandoned thesis 'The Ideality of the Literary Object' may be matter for surmise. It remains only to give to the writer of 'The Time of a Thesis: Punctuations' [1980] the last word:

> And as is often, as is always the case, it is the friendly advice of this or that person among those present here, before or behind me, it is others, always others, who effected in me a decision I could not have come to alone. For not only am I not sure, as I never am, of being right in taking this step, I am not sure that I see in all clarity what led me to do so. Perhaps because I was beginning to know only too well not where I was going but where I was, not where I had arrived but where I stopped.
>
> (D: 2004, EU p. 127)[25]

Notes

1.1 On time and movement

1 See Aristotle, *Physics*, a new translation by Robin Waterfield, with an introduction and notes by David Bostock (Oxford: Oxford University Press 1996) and Edward Hussey (ed.) *Aristotle's Physics: Books III and IV,* Clarendon Aristotle series, general editor J. L. Ackrill (Oxford: Clarendon Press 1983).

2 See the lectures from 1921–22, GA 61, Martin Heidegger, *Phenomenological Interpretations of Aristotle: Initiation into Phenomenological Research*, trans. Richard Rojcewicz (Bloomington IN: Indiana University Press, 2001); and from 1923–24, GA 17, Martin Heidegger, *Introduction to Phenomenological Research* [1994] trans. Daniel O. Dahlstrom (Bloomington IN: Indiana University Press, 2005). See especially the former:

> 'Time' is to be understood here neither as a framework for ordering things, a dimension of order, nor as the (specifically formal) character of the connections among historical events, but as a specific *mode of movedness* in the sense of a character that not only makes movedness possible, releasing it from within, but co-constitutes it as itself moving in an autonomously factical way.
>
> (GA 61, p. 103)

For Aristotle, *De Anima*, see Hugh Lawson Tancred (ed.) (Harmondsworth, UK: Penguin 1986).

3 For the Freudian topographies, see 'The Unconscious', section 2, and the beginning of 'The Ego and the Id', in Sigmund Freud, *On Meta-psychology and the Theory of Psychoanalysis*, Penguin Freud Library 11 (Harmondsworth, UK: Penguin 1984), pp. 174–78, and pp. 339–408. In the former Freud writes:

> By accepting the existence of these two (or three) psychical systems, psychoanalysis has departed a step further from the descriptive 'psychology of consciousness' and has raised new problems and acquired a new content. Up until now, it has differed from that psychology mainly by reason of its dynamic view of mental processes; now in addition it seems to take account of psychical topography as well, and to indicate in respect of any given mental act within what system or between what systems it takes place.
>
> (PFL pp. 175–76)

In the subsequent essay the new topography of id, ego and superego is introduced, and Freud writes:

> we have arrived at the term or concept of the unconscious along another path, by considering certain experiences in which mental dynamics play a part. We

> have found – that is, we have been obliged to assume – that very powerful mental processes or ideas exist (and here a quantitative or economic factor comes into question for the first time) which can produce all the effects in mental life that ordinary ideas do (including effects that can in their turn become conscious as ideas) though they themselves do not become conscious.
>
> (p. 352)

This quantitative factor is contrasted by Freud to a previously invoked notion of the qualitative:

> To put it once more, in a different way, psychoanalysis cannot situate the essence of the psychical in consciousness, but is obliged to regard consciousness as a quality of the psychical which may be present in addition to other qualities or may be absent.
>
> (p. 351)

The priority here given to the quality of consciousness over quantative forces informs Derrida's reading of Kant in *The Truth in Painting* [1978] trans. Geoffrey Bennington and Ian McLeod (Chicago: University of Chicago Press 1987), both of Freud and of course of Kant's reversal of the order between quality and quantity, in the discussion of the categories, in the *Critique of Judgment* (1791). See especially pp. 67–77. The differences between descriptive, dynamic and economic accounts of psychical structures could be deployed to good effect to distinguish between aspects of Husserl's analyses.

4 For reworkings of the encounter with Aristotle, see Heidegger, *Basic Problems of Phenomenology* [1927] trans. Albert Hofstadter (Bloomington IN: Indiana University Press 1988) especially section 19, where the names of Aristotle and of Augustine are conjoined. Heidegger had lectured on Augustine, on time, in the 1924 lecture *The Concept of Time*, trans. Will McNeill (Cambridge: Cambridge University Press 1998).

5 For notes towards a genealogy of the *psyche*, see Jacques Derrida, *Psyche: inventions de l'autre* [1990], 2nd edn, 2 vols (Paris: Editions Galilée 1998, 2003).

6 Jacques Derrida: '*Ousia* and *Gramme*: Note on a Note from *Being and Time*', in Derrida, *Margins of Philosophy* [1972] trans. with notes by Alan Bass (Brighton, UK: Harvester Press 1982), pp. 31–66. The exergues from Hegel with which the first essay in this collection opens, make Hegel the principal interlocutor. In this essay Derrida addresses himself to Aristotle's *Physics 4*, quoted above, and remarks:

> There is time only in the extent to which movement has number, but time, in the rigorous sense is neither movement nor number. It lets itself be numbered only insofar as it has a relation to movement according to the before and after.
>
> (D: 1972, MP p. 59)

The hypothesis pursued here is that there are more articulations of time to be added to and taken away from this 'before and after', not least the time of the *apriori*, which doubles the 'before and after' of empirical events with a 'before and after' of logical necessities. The last paragraph of the essay reads:

> Such a *différance* would at once, again, give us to think a writing without presence and without absence, without history, without cause, without *archia*, without *telos*, a writing that absolutely upsets all dialectics, all theology, all teleology, all ontology. A writing exceeding everything that the history of metaphysics has comprehended in the form of the Aristotelian *gramme*, in its point, in its line, in its circle, in its time and in its space.
>
> (p. 67)

These additional articulations of time are given in the non-regular elliptical movements marked up in relation to reading Husserl, in the distinct movements of repetition and iteration, in the movements of givenness, and the divagations induced by accelerations and decelerations of pace of this 'at once, again', which signals the originary asynchrony and achrony, the 'non-contemporaneity with itself of the living present' (p. xix) brought to the fore in Jacques Derrida, *Specters of Marx: The State of the Debt, The Work of Mourning and the New International* [1993] trans. Peggy Kamuf (London: Verso 1994).

7 Martin Heidegger, *The Metaphysical Foundations of Logic* [1928] trans. Michael Heim (Bloomington IN: Indiana University Press 1984) section 12.

8 See the essay 'Hegel's Concept of Experience' in *Holzwege/Off the Beaten Track* [1950] trans. Kenneth Haynes and Julian Young (Cambridge: Cambridge University Press 2002); and the volume of lectures from 1930–31, GA 32, *Hegel's Phenomenology of Spirit* [1988] trans. Parvis Emad and Kenneth Maly (Bloomington IN: Indiana University Press 1988). For an excellent discussion of the relation between Heidegger and Hegel on the thinking of time, see Karin de Boer, *Thinking in the Light of Time: Heidegger's Encounter with Hegel* [1997] (Albany NY: SUNY Press 2000).

9 For a discussion of this see Joanna Hodge, 'Phenomenologies of Faith and Hope: Heidegger on Religion', *Journal of the British Society for Phenomenology*, vol. 37, no 1, January 2006, special issue on Religion, ed. Jim Urpeth.

10 See G. W. F. Hegel, *Philosophy of Mind*, being part 3 of *The Encyclopaedia of the Philosophical Sciences* [1830] trans. William Wallace and A. V. Miller (Oxford: Clarendon Press 1971) p. 315; and Aristotle, *Metaphysics* lamda, 7, trans. Hugh Lawson-Tancred (Harmondsworth, UK: Penguin 1998) pp. 373–74.

11 See Jacques Derrida, *L'animal que donc je suis*, texte établie par Marie-Louise Mallet (Paris: Editions Galilée 2006). Part of an early version of this (pp. 15–77) is translated by David Wills as Jacques Derrida, 'The Animal Which therefore I Am (More to Follow)', *Critical Inquiry*, vol. 28, no. 2, pp. 369–418, winter 2002 (Chicago IL: University of Chicago Press). The former is cited as D: 2006, AQD; the latter as D: 1997, AA.

12 For Jean-Luc Nancy on the experience of a finitude of time, see especially Nancy, *The Experience of Freedom* [1988] trans. Bridget McDonald (Stanford CA: Stanford University Press 1993); *A Finite Thinking*, trans. Simon Sparks (Stanford CA: Stanford University Press 2003); and *Being Singular Plural* [1996] trans. Robert D. Richardson and Annie E. O'Byrne (Stanford CA: Stanford University Press 2000). See also the essay by Christopher Fynsk, 'The Experience of Finitude', which serves as an introduction to Jean-Luc Nancy, *The Inoperative Community* [1986] trans. Peter Connor *et al.* (Minneapolis MN: University of Minnesota Press 1991).

13 Jacques Derrida, *The Problem of Genesis in Husserl's Philosophy* [1990] trans. Marion Hobson (Chicago: University of Chicago Press, 2000); and Jacques Derrida, *On Touching: Jean-Luc Nancy* [2000] trans. Christine Irizarry (Stanford CA: Stanford University Press 2005).

14 See Jacques Derrida, *Of Grammatology* [1967] trans. Gayatri Chakravorty Spivak (Baltimore MD: Johns Hopkins University Press 1974, 1976). For a remarkable reading of Derrida's readings of Husserl which arrives at a diagnosis of the thinking of trace and of supplementarity as resulting from those readings, see Joshua Kates, *Essential History: Jacques Derrida and the Development of Deconstruction* (Evanston IL: Northwestern University Press 2005).

15 See Edmund Husserl, *The Crisis of the European Sciences and Transcendental Phenomenology* [1970] trans. David Carr (Evanston IL: Northwestern University Press 1970).

16 Heidegger remarks the importance of this distinction in the course of the introduction to his 1922 lectures, *Phenomenology of Religious Life* (GA 60) trans. Matthias Fritsch and Jennifer Anna Gosetti-Ferencei (Bloomington IN: Indiana University Press 2002).

17 Edmund Husserl, with Ludwig Landgrebe, *Experience and Judgment: Investigations in a Genealogy of Logic* [1938] trans. James Churchill and Karl Ameriks (Evanston IL: Northwestern University Press 1973).

18 Jacques Derrida, *Speech and Phenomenon: An Introduction to the Problem of Signs in Husserl's Phenomenology* [1967] trans. David Allison (Evanston IL: Northwestern University Press 1974).

19 For the introduction of the term '*Nachträglichkeit*', see Freud, *Case Histories II*: 'From the History of an Infantile Neurosis: The "Wolf Man"', trans. James Strachey, general editor Angela Richards, Penguin Freud Library vol. 9 (Harmondsworth, UK: Penguin 1979) the footnote to IV, The Dream and the Primal Scene, p. 278, where the term is translated as 'deferred action', and IX: Recapitulation and Problems, p. 349. For its importance to Lacan, as a splitting of time, see the 1953 Rome Discourse, 'The Function and the Field of Speech and Language', in Jacques Lacan, *Ecrits*, trans. Alan Sheridan (London: Tavistock Press 1977) p. 48.

20 Edmund Husserl: *Ideas: A General Introduction to Phenomenology* [1913] trans. W. Boyce Gibson (New York: George Allen and Unwin 1931), commonly known as *Ideas One*, and *Ideas Pertaining to a Pure Phenomenology and to a Phenomenological Philosophy*, trans. Richard Rojcewicz and Andre Schuwer (Dordrecht: Kluwer Academic Publishers 1989).

21 Edmund Husserl, *Analyses Concerning Passive and Active Synthesis: Lectures on Transcendental Logic* [1920/21] trans. Anthony Steinbock (Dordrecht: Kluwer Academic Publishers 2001).

22 See Jacques Derrida, *Psyche: inventions de l'autre* (1987–2003).

23 For what in effect is a long footnote to this remark, see Joanna Hodge, 'Husserl, Freud, *à suivre*: Derrida on Time', *Journal of the British Society for Phenomenology*, vol. 36, no. 2, May 2005, special issue on Husserl and Derrida, ed. Gary Banham, pp. 188–207. For the movement of interruption see Part II below.

24 Jacques Derrida: 'To Speculate – on "Freud"', in *The Post Card: From Socrates to Freud and Beyond* [1980] trans Alan Bass (Chicago: University of Chicago Press 1987) pp. 257–409. Derrida quotes Freud from 'Beyond the Pleasure Principle' in Sigmund Freud, *On Metapsychology: The Theory of Psychoanalysis*, trans. James Strachey, ed. Angela Roberts (Harmondsworth, UK: Penguin 1984): 'the pleasure and unpleasure series indicates a change (*Änderung*) in the magnitude of the cathexis within a given unit of time' and remarks:

> The 'unit of time' (*Zeiteinheit*) is not cut out from within the homogenous element of a form of the senses. This must be remarked, without engaging ourselves here in this immense problem. I have attempted to situate it elsewhere (in 'Freud and the Scene of Writing'), and it demands a systematic 'explication' between shall we say Freud and at least, for example, Aristotle, Kant, Hegel, Husserl, Heidegger: on the question of time.
>
> (D: 1980, PC p. 407)

In 'Freud and the Scene of Writing', in Derrida, *Writing and Difference* [1967] trans. Alan Bass (London: Routledge and Kegan Paul 1978) pp. 196–231, Derrida observes: 'Time is the economy of a system of writing' (p. 226), a remark which I have attempted to test here by setting out the trajectory of a reading of Husserl, and the resulting proliferation of phenomenological enquiries emerging from responses to his writings. 'Economy' is a word for which Derrida subsequently finds other forms of expression, in order to analyse the effects of a dis-equilibrium between the scopes of general and restricted economies. Thus analysis of Freud's topographies gives way to the inventions of parergonality, topolitology, pragrammatology and limitrophy. In a footnote to this text, Derrida remarks: 'The concept of originary *différance* and orginary delay (*retard*) were imposed on us by a reading of Husserl' (D: 1967, WD p. 329).

25 The relevant essays here are Jacques Lacan, 'Seminar on "The Purloined Letter"', Yale French Studies, no. 48, *French Freud: Structural Studies in Psychoanalysis* (1972) pp. 39–72; and Jacques Derrida, '*Le facteur de la vérité*', in Jacques Derrida, *The Post Card: From Socrates to Freud and Beyond* [1980], pp. 412–96. For the connection between the *envoi*, the *dessein* and the *destin* (the despatch, the design and the destination), see the discussion appended to it under the title 'Du tout', pp. 497–521, esp. p. 513.

26 See Maurice Blanchot, *The Step (not) Beyond* [1973] trans. Lycette Nelson (Albany NY: SUNY Press 1992). For the interaction between Levinas and Blanchot see also Will Large, *Emmanuel Levinas and Maurice Blanchot: Ethics and the Ambiguity of Writing* (Manchester: Clinamen Press 2005).

27 See Jean-Luc Nancy: 'Elliptical Sense' (1987) trans. Peter Connor, in David Wood (ed.) *Derrida: A Critical Reader* (Oxford: Blackwell 1992) pp. 36–51:

> Thus the transcendental experience of writing is not the 'transcendental experience' described by Husserl. Husserl's transcendental experience was a 'pure experience' which reduced and purified empiricalness. Here experience is impure, and this is indeed why neither the concept of experience, if this supposes at least some elaboration on the experiential level nor the concept of the transcendental (which is always the concept of a knowledge which, as a condition of possibility, is *a priori*) can be considered as adequate.
>
> (p. 48)

This circumscribes the transcendental, in relation to the possible, the actual and a certain concept of the necessary, which is here to be opened out towards a thinking of a concept of the transcendental as conditions for the probable and improbable, for the 'perhaps', of what may occur, as a modality less restricted in scope than that of this necessity. This latter seems to me to set out the logic of this impure experience. See also Joanna Hodge, 'Phenomenology to Come: Derrida's Ellipses' in Martin McQuillan (ed.) *The Politics of Deconstruction: Jacques Derrida and the Other of Philosophy* (London: Pluto Press, 2007).

28 Jacques Derrida, 'Form and Meaning: A Note on the Phenomenology of Language', in Derrida, *Margins of Philosophy* (1982) [1972] trans. Alan Bass, pp. 155–73.

29 Jacques Derrida, '"Genesis and Structure" and Phenomenology' [1959] in Derrida, *Writing and Difference* (1978) [1967] trans. Alan Bass, pp. 154–67.

30 See Edmund Husserl, *Ideas: A General Introduction to Phenomenology* [1913].

31 See '*Différance*' [1968] trans. David B. Allison, in Jacques Derrida, *Speech and Phenomena and Other Essays on Husserl's Theory of Signs* [1967] pp. 129–60, which translates the essay as printed, in the *Bulletin de la société française de philosophie*, vol. LXII (1968) with a prefatory remark, and '*Différance*' [1972] trans. Alan Bass, in Jacques Derrida, *Margins of Philosophy* (1982) pp. 1–27, which prints it as delivered.

32 It is the inestimable service of Len Lawlor's *Derrida and Husserl: The Basic Problem of Phenomenology* (Bloomington IN: Indiana University Press 2002) to have drawn attention to the continuing impact on Derrida's thinking of Husserl's concept of genesis.

33 Augustine, *Confessions*, trans. Henry Chadwick (Oxford: Oxford University Press 1997).

34 As discussed by Derrida in the second part of *Of Grammatology* [1967].

35 See Immanuel Kant, *Critique of Pure Reason* [1781, 1787] trans. Norman Kemp Smith (London: Macmillan 1929). The introduction of further antinomies in the *Critique of Practical Reason*, concerning virtue and happiness, and in the *Critique of Judgment* concerning taste and efficient and final causes, further complicates matters. The question whether there are one or more forms of antinomy is not often addressed in the context of all seven instances adduced by Kant. Derrida discusses antinomy in 'Privileges', the introduction to Jacques Derrida: *Who's Afraid of Philosophy: Right to Philosophy I* [1990] trans. Jan Plug (Stanford CA: Stanford University Press 2002) and deploys the term in his essay 'The Antinomies of the Philosophical Profession: An Introduction', trans. Jan Plug, in *Eyes of the University (Right to Philosophy 2)* [1990] (Stanford: Stanford University Press 2004). He also uses the term in 'Typewriter Ribbon: Limited Ink. (2)' to be discussed in Part V of these enquiries; see Jacques Derrida, *Without Alibi*, ed. Peggy Kamuf (Stanford CA: Stanford University Press 2003) pp. 71–160.

36 See Edmund Husserl, *On the Phenomenology of the Consciousness of Internal Time (1893–1917)* (HUA 10) trans. John Barnet Brough (Dordrecht: Kluwer Academic Publishers

1991) for the distinction between dimensions of intentionality. See also Rudolf Bernet, 'An Intentionality without Subject or Object', *Man and World*, vol. 27 (1994) pp. 231–55.

37 See Jacques Derrida, 'Violence and Metaphysics: An Essay on the Thought of Emmanuel Levinas', in *Writing and Difference* (1978) pp. 79–153.

38 For an excellent discussion of Derrida's encounter with Foucault, see Marion Hobson, *Jacques Derrida: Opening Lines* (London: Routledge 1998).

39 Jacques Derrida, 'Of an Apocalyptic Tone Recently Raised in Philosophy' is to be found in Harold Coward and Toby Foshay (eds) trans. John P. Leavey Jr, *Derrida and Negative Theology* (Albany NY: SUNY Press 1992). It is also to be found in Peter Fenves (ed.) *Raising the Tone of Philosophy: Late Essays by Immanuel Kant, Transformative Critique by Jacques Derrida* (Baltimore MD: Johns Hopkins University Press 1993) which conveniently also prints the essay by Kant. See also Peter Fenves' study, *A Peculiar Fate: Metaphysics and World-History in Kant* (Ithaca NY: Cornell University Press 1991). The paper 'Of an Apocalyptic Tone' was presented at the first Cérisy conference, dedicated to the work of Derrida, entitled 'The Ends of Man', the proceedings of which were edited by Jean-Luc Nancy and Philippe Lacoue-Labarthe: *Les Fins de L'Homme* (Paris: Editions Galilée 1983).

40 For Heidegger's analysis of the necessity for this move, see 'Letter on Humanism', in Martin Heidegger, *Pathmarks* (1976, GA 9) trans. William McNeill (Cambridge: Cambridge University Press 1998) pp. 239–76.

41 See Jacques Derrida, 'Heidegger's Ear: of *Philopolemology* (*Geschlecht* IV)', in John Sallis (ed.) *Reading Heidegger: Commemorations* (Bloomington IN: Indiana University Press 1993) pp. 163–220, where there is a continuation of the engagement with Heidegger from *Of Spirit* (1987). See also the first two *Geschlecht* papers, Jacques Derrida, '*Geschlecht*: Sexual Difference, Ontological Difference' (1983) in *A Derrida Reader: Between the Blinds*, ed. Peggy Kamuf (London and New York: Harvester Wheatsheaf 1991); and Jacques Derrida, '*Geschlecht II*: Heidegger's Hand', trans. John P Leavey Jr, in John Sallis (ed.) *Deconstruction and Philosophy: The Texts of Jacques Derrida* (Chicago: University of Chicago Press 1987) pp. 161–96.

42 For an indispensable reading of Derrida's responses to Husserl and Heidegger, see Paola Marrati, *Genesis and Trace: Derrida Reading Husserl and Heidegger* [1998] trans. Simon Sparks (Stanford: Stanford University Press 2005), which provides excellent detailed readings of Derrida's readings of these two and emphasises the question of transcendental historicity. This notion of transcendental historicity is both the inheritor of the notion of intentionality, as transformed by Heidegger and Levinas, and the transition to the specifically Derridean notion of textuality as autobiography.

43 Derrida reaffirms the difficulty of releasing Kierkegaard's analyses from their Christian inscription in the final pages of his interview with Giovanna Borradori, 'Autoimmunity: Real and Symbolic Suicides: A Dialogue with Jacques Derrida', in Borradori (ed.) *Philosophy in a Time of Terror: Dialogue with Jürgen Habermas and Jacques Derrida* (2003) (Chicago: University of Chicago Press 2003) pp. 85–136.

44 For Jean-Luc Marion's readings of Husserl, see *Reduction and Givenness: Investigations of Husserl, Heidegger and Phenomenology*, trans. Thomas A. Carlson (Evanston IL: Northwestern University Press 1998); and *Being Given: Toward a Phenomenology of Givenness* [1997] trans. Jeffrey L. Kosky (Stanford CA: Stanford University Press 2002). For the debates with Derrida, see John D. Caputo and Michael J. Scanlon (eds) *God, the Gift and Postmodernism* (Bloomington IN: Indiana University Press 1999). For discussion of Marion, see Ian Leask and Eoin Cassidy, *Givenness and God: Questions of Jean-Luc Marion* (New York: Fordham University Press 2005).

1.2 'Aphorism countertime'

1 See Jacques Derrida, 'Aphorism Countertime', in Derek Attridge (ed.) *Jacques Derrida: Acts of Literature* (New York and London: Routledge 1992) pp. 414–33.

2 For the importance to Derrida of a notion of tradition, transmission, and legacy see Michael Naas, *Taking on the Tradition: Jacques Derrida and the Legacies of Deconstruction* (Stanford CA: Stanford University Press 2003).

3 See Jacques Derrida, 'Living On – Border Lines', in Harold Bloom, Paul de Man, Jacques Derrida, Geoffrey Hartman and J. Hillis Miller, *Deconstruction and Criticism* (New York: Continuum 1992) pp. 73–176; and Edmund Husserl, *The Phenomenology of Internal Time Consciousness* [1928] ed. Martin Heidegger, trans. James Churchill (Bloomington IN: Indiana University Press 1973).

4 See Edmund Husserl, *Cartesian Meditations* [1931] trans. Dorion Cairns (The Hague: Martinus Nijhoff 1960); and G. Leibniz, 'The Monadology', in Georg Leibniz, *Discourse on Metaphysics and Other Essays*, trans. Daniel Garber and Roger Ariew (Indianapolis IN: Hackett 1991).

5 Eugen Fink, *Sixth Cartesian Meditation: The Idea of a Transcendental Theory of Method*, with textual notations by Edmund Husserl [1988] trans. Ronald Bruzina (Bloomington IN: Indiana University Press 1995).

6 Jacques Derrida, *The Problem of Genesis in Husserl's Philosophy* [1990] trans. Marion Hobson (Chicago: University of Chicago Press 2001).

7 Edmund Husserl, *On the Phenomenology of the Consciousness of Internal Time* [1966] HUA 10, HUE 4, trans. John Barnet Brough (Dordrecht: Kluwer Academic Publishers 1991). The first part of this was translated as Edmund Husserl, *The Phenomenology of Internal Time Consciousness* [1928] ed. Martin Heidegger, trans. James Churchill (Bloomington IN: Indiana University Press 1973).

8 Edmund Husserl, *Die 'Bernauer Manuskripte' über das Zeitbewusstsein 1917–1918*, ed. Rudolf Bernet and Dieter Lohmar (Dordrecht: Kluwer Academic Publishers 2001). See also Toine Kortooms, *Phenomenology of Time: Edmund Husserl's Analyses of Time Consciousness* (Dordrecht: Kluwer Academic Press 2002); and see my review, Hodge, 'Excesses of Subtlety: The Current Reception of Edmund Husserl', *Journal of the British Society for Phenomenology*, vol. 35, no. 2, May 2004, pp. 208–13.

9 Edmund Husserl, *Ideas Pertaining to a Pure Phenomenology and to a Phenomenological Philosophy, Second Book, Studies in the Phenomenology of Constitution* (known as *Ideas Two*, 1952) trans. Richard Rojcewicz and Andre Schuwer (Dordrecht: Kluwer Academic Publishers 1989).

10 See Maurice Merleau-Ponty, *The Phenomenology of Perception* [1945] trans. Colin Smith (London: Routledge and Kegan Paul 1962); and Martin Heidegger, *History of the Concept of Time* (1925, GA 20) trans. Theodore Kisiel (Bloomington IN: Indiana University Press 1988).

11 Len Lawlor draws attention to the importance of this review in his *Derrida and Husserl: The Basic Problem of Phenomenology*; see Jacques Derrida, 'Review of Edmund Husserl, *Phänomenologische Psychologie. Vorlesungen Sommersemester* (1925, HUA 9) in *Les études philosophiques*, vol. 18, no. 2 (1963) pp. 203–6.

12 For analyses of this, see Sean Gaston, *Derrida and Disinterest* (London: Continuum 2005).

13 Jacques Derrida, *Speech and Phenomena and Other Essays on Husserl's Theory of Signs*, trans. David Allison (Evanston IL: Northwestern University Press, 1973). I shall adopt the convention of referring to this as *Speech and Phenomenon*, when only the text of 1967 is in question.

14 See Jacques Derrida, 'Freud and the Scene of Writing', in Derrida, *Writing and Difference* [1967] trans. Alan Bass (London: Routledge 1978), footnote, p. 329, in which is to be found the remark: 'The concepts of originary *differance* and originary delay (*retard*) were imposed on us by a reading of Husserl.'

15 See Sigmund Freud, *On Metapsychology: The Theory of Psychoanalysis*, Penguin Freud Library vol. 11, trans. James Strachey, ed. Angela Richards (Harmondsworth, UK: Penguin 1984).

16 See Jacques Derrida, '*Mes Chances*: A Rendezvous with Some Epicurean Stereophonies', in Joseph H. Smith and William Kerrigan (eds) *Taking Chances: Derrida, Psychoanalysis and Literature* (Baltimore MD and London: Johns Hopkins University Press 1984) pp. 1–32.

17 See Paul Ricoeur, *Husserl: An Analysis of his Phenomenology*, trans. Edward G. Ballard and Lester E. Embree (Evanston IL: Northwestern University Press 1967).
18 See Emmanuel Levinas, *Time and the Other and Other Essays* [1948] trans. Richard A. Cohen (Pittsburgh PA: Duquesne University Press 1987) pp. 38–94.
19 See Paul Ricoeur, *Husserl: An Analysis of his Phenomenology*, op. cit.
20 See Immanuel Kant, *Critique of Pure* Reason, trans. Norman Kemp Smith (London: Macmillan 1929) A426/B454-A444/B471, pp. 396–409.
21 See Jacques Derrida, *The Gift of Death* [1992] trans. David Wills (Chicago: University of Chicago Press 1995).
22 See Gaston Bachelard, *The Dialectics of Duration* [1950] trans. Mary McAllester Jones (Manchester: Clinamen Press 2000).
23 For Jacques Derrida, '"Genesis and Structure" and Phenomenology' [1959], see Jacques Derrida, *Writing and Difference*, trans. Alan Bass (London: Routledge 1987) pp. 154–68.
24 See Jean-François Lyotard, *The Differend: Phrases in Dispute* [1983] trans. George van Abbeela (Minneapolis MN: University of Minnesota Press 1988).
25 See Jacques Derrida, 'Lyotard and Us' [1998] in Jacques Derrida, *The Work of Mourning*, trans. and eds Pascale Anne Brault and Michael Naas (Chicago: University of Chicago Press 2002) pp. 211–41.
26 For Jacques Derrida, 'The Ends of Man' [1968] see Jacques Derrida, *Margins of Philosophy* [1972] trans. Alan Bass (Brighton, UK: Harvester Wheatsheaf 1982).
27 For discussions of this, see James Dodd, 'Reading Husserl's Time Diagrams from 1917/1918', *Husserl Studies*, vol. 21, no. 2 (2005) pp. 111–37; and Dan Zahavi, 'Time and Consciousness in the Bernau Manuscripts', *Husserl Studies*, vol. 20, no. 2 (2004) pp. 99–118.

2.1 The sense of genesis and the genesis of sense

1 Jacques Derrida, 'And etc.', in Nicholas Royle (ed.) *Deconstruction: A Critical Reader* (London: Macmillan 2000).
2 Jacques Derrida, *The Problem of Genesis in Husserl's Phenomenology* [1991] trans. Marion Hobson, (Chicago: University of Chicago Press 2001) subsequently referred to as (D: PG).
3 Immanuel Kant, *Critique of Pure Reason* [1781, 1787] trans. Norman Kemp Smith (London: Macmillan 1929, 1976); see its introduction at A295, B352, pp. 298f., where Kant distinguishes between optical, logical and transcendental illusion. The dialectic, on the model of the analytic of elements and principles, first explores illusory concepts, and then illusory or dialectical inferences.
4 Edmund Husserl and Ludwig Landgrebe, *Experience and Judgment: Towards a Genealogy for Logic* [1938] trans. James Churchill and Karl Ameriks (Evanston IL: Northwestern University Press 1968); and Edmund Husserl, *Analyses Concerning Passive and Active Synthesis: Introduction to Transcendental Logic* (HUE 9) trans. Antony Steinbock (Dordrecht: Kluwer Academic Publishers 2002). The latter but not the former is deemed to be authoritative text, although the former was intended for publication by Husserl while the latter is a compilation from a number of lecture series given in the 1920s. Husserl wrote *Formal and Transcendental Logic* [1928] trans. Dorion Cairns (Martinus Nijhoff: The Hague 1969) as an introduction to the former.
5 Edmund Husserl, *Logical Investigations* (1900–) trans. J. F. Findlay (London: Routledge 1970).
6 Jacques Derrida, *Speech and Phenomena and Other Essays on Husserl's Theory of Signs* [1967] trans. David Allison (Evanston IL: Northwestern University Press 1973), subsequently referred to as (D: SP).
7 Edmund Husserl, *Die 'BernauerManuskripte' über das Zeitbewusstsein 1917/1918*, eds Rudolf Bernet and Dieter Lohmar (Dordrecht: Kluwer Academic Publishers 2001). For some thoughts on the *Bernauer Manuscripts*, see Joanna Hodge, 'Excesses of Subtlety: The Current Reception of Edmund Husserl', *Journal of the British Society for Phenomenology*, vol. 35, no. 2, May 2004, pp. 208–13.

8 Jacques Derrida, *Speech and Phenomena and Other Essays on Husserl's Theory of Signs* [1967].
9 Aristotle, *De Anima*, trans. Hugh Tancred Lawson (Harmondsworth, UK: Penguin 1998).
10 For some parallel questions to Nancy on touching, see Joanna Hodge, 'Why Aesthetics Might Be Several: Nancy on Art and Technicity', *Angelaki: Journal of the Theoretical Humanities*, vol. 7, no. 1, April 2002, pp. 53–68.
11 Edmund Husserl, *Ideas: A General Introduction to Pure Phenomenology* [1913] trans. W. B. Boyce Gibson (London: Collier Macmillan 1974), known as *Ideas One*.
12 The importance of the reading of Rousseau, in *Of Grammatology* (1967) for an understanding of Derrida's reading of Husserl is brought out by Bernard Stiegler in his study *Technics and Time: The Fault of Epimetheus*, vol. 1 [1994] ch. 2, pp. 82–133, trans. Richard Beardsworth and George Collins (Stanford CA: Stanford University Press 1998).
13 Jacques Derrida, *Introduction to Husserl's 'Origin of Geometry'* [1962] trans. John P. Leavey Jr (Brighton, UK: Harvester 1978), subsequently referred to as (D: 1962, HOG).
14 See Edmund Husserl, *The Crisis of the European Sciences and Transcendental Phenomenology* [1952] trans. David Carr (Evanston IL: Northwestern University Press 1970).
15 See also Edmund Husserl, *Formal and Transcendental Logic* [1928] trans. Dorion Cairns (The Hague: Martinus Nijhoff 1969).
16 See Derrida, *Introduction to Husserl's 'Origin of Geometry'*, and also Paul Ricoeur, *Husserl: An Analysis of his Phenomenology*, trans. Edward G. Ballard and Lester E. Embree (Evanston IL: Northwestern University Press 1967).

2.2 Delay, difference, *differance*

1 Jacques Derrida, *Given Time I: Counterfeit Money* [1991] trans. Peggy Kamuf (Chicago: University of Chicago Press 1992), subsequently referred to as D: 1991, GT.
2 Jacques Derrida, *The Post Card: From Socrates to Freud and Beyond* [1980] trans. Alan Bass (Chicago: University of Chicago Press 1987) cited as D: 1980, PC.
3 For Lacoue-Labarthe, see Philippe Lacoue-Labarthe, *Typography: Mimesis, Philosophy, Politics*, ed. Christopher Fynsk, with an introduction 'Desistances' by Jacques Derrida (Cambridge MA: Harvard University Press 1989).
4 'Différance' is published in English in two versions, one in Jacques Derrida, *Margins of Philosophy* [1972] trans. Alan Bass (Brighton, UK: Harvester Press 1982) pp. 3–27, gives the lecture as delivered; the other, in Jacques Derrida, *Speech and Phenomena and Other Essays on Husserl's Theory of Signs*, trans. and ed. David Allison (Evanston IL: Northwestern University Press 1973) pp. 129–60, gives that same text with the addition of the explanatory preface, added for its original publication in the *Bulletin de la société française de philosophie*, vol. LXII, no. 3, July-September 1968).
5 See Edmund Husserl, *Ideas: A General Introduction to Phenomenology* [1913] trans. W. Boyce Gibson (New York: George Allen and Unwin 1931) commonly known as *Ideas One*. See also Jean Luc Marion, *Reduction and Giveness: Investigations of Husserl, Heidegger and Phenomenology* [1989] trans. Thomas A. Carlson (Evanston IL: Northwestern University Press 1998).
6 See Jacques Derrida, 'Lyotard and Us' [1998] in Derrida, trans. and eds Pascale Anne Brault and Michael Naas, *The Work of Mourning* (Chicago: University of Chicago Press 2002).
7 Immanuel Kant, *The Critique of Pure Reason* [1781, 1787] trans. Norman Kemp Smith (London: Macmillan 1929).

3.1 'A time that has always already run out'

1 See Jacques Derrida, *Adieu: à Emmanuel Levinas*, trans. Pascale Anne Brault and Michael Naas (Stanford CA: Stanford University Press 1998) pp. 1–13.
2 See Emmanuel Levinas, 'Bad Conscience and the Inexorable', first published in the review, *Exercices de la patience 2* (Paris: Editions Obsidian 1981) translated in *Of God Who*

Comes to Mind [1986] by Bettina Bergo (Stanford CA: Stanford University Press 1998) pp. 172–77. The citation at the opening of this piece from Maurice Blanchot, *The Writing of the Disaster* [1980] trans. Anne Smock (Lincoln NE and London: University of Nebraska Press 1986):

> The I responsible for the other, an I without an I is fragility itself, to the point of being put in question through and through *qua* I, without identity, responsible for him to whom he cannot give a response, a responding that is not a question, a question that refers to the other without so much as waiting for a response from him. The Other does not respond.
>
> (B: WD, p. 119)

3 See Blanchot, op. cit.
4 Blanchot here is responding to Levinas' analyses and criticisms of Husserl on otherness, beginning already in the lectures by Emmanuel Levinas, *Time and the Other* [1948] trans. Richard Cohen (Pittsburgh PA: Duquesne University Press 1969), and the analyses of the phenomenology of both Husserl and Heidegger commenced in *Existence and Existents* [1947] trans. Alphonso Lingis (Dordrecht: Kluwer 1978).
5 For some discussions of this, see excellent essays by Paul Davies, 'A Fine Risk: Reading Blanchot Reading Levinas', in Robert Bernasconi and Simon Critchley (eds) *Re-reading Levinas* (London: Athlone Press 1991); and 'Sincerity and the End of Theodicy: Three Remarks on Levinas and Kant', in Simon Critchley and Robert Bernasconi (eds) *The Cambridge Companion to Levinas* (Cambridge: Cambridge University Press 2002). See also the series of studies by John Llewelyn, *Derrida: On the Threshold of Sense* (London: Macmillan 1986); *The Hypocritical Imagination: Between Kant and Levinas* (London: Routledge 2000); and *Appositions of Jacques Derrida and Emmanuel Levinas* (Bloomington IN: Indiana University Press 2002). See also Will Large, *Emmanuel Levinas and Maurice Blanchot: Ethics and the Ambiguity of Writing* (Manchester: Clinamen 2005), the writings of Jean-Luc Nancy and of Jean-François Lyotard, especially J-L. Nancy, *The Experience of Freedom* [1988] trans. Bridget McDonald (Stanford CA: Stanford University Press 1993) on a similar relation emergent in the process of reading Heidegger reading Kant; and J-F. Lyotard, *The Differend: Phrases in Dispute* [1983] trans. Georges van den Abbeele (Minneapolis MN: University of Minnesota Press 1988). The latter considers a deployment of Greek sophistical paradox in the articulation of Holocaust denial.
6 On friendship, see Maurice Blanchot, *Friendship* [1971] trans. Elizabeth Rottenberg (Stanford CA: Stanford University Press 1997); and Jacques Derrida, *Politics of Friendship* [1994] trans. George Collins (London: Verso 1997).
7 For a succinct statement of the position, see Edmund Husserl, *Idea of Phenomenology: Five Lectures from 1907*, HUA 2, HUE 8, trans. Lee Hardy (Dordrecht: Kluwer 1999). For the discussion of inter-subjectivity see Edmund Husserl, *Cartesian Mediations* [1931] HUA 1 trans. Dorion Cairns (The Hague: Martinus Nijhoff 1960) especially the Fifth Meditation.
8 See Martin Heidegger, *Being and Time* [1927] trans. John Macquarrie and Edward Robinson (Oxford: Blackwell 1962), introduction, sections 1–7.
9 See Martin Heidegger, *Of the Event: Contributions to Philosophy* [1989] trans. Parvis Emad (Bloomington IN: Indiana University Press 1999).
10 On Blanchot's reading of Heidegger, see Herman Rapaport, *Heidegger and Derrida: Reflections on Time and Language* (Lincoln NE and London: University of Nebraska Press 1989); see also Leslie Hill, *Blanchot: Extreme Contemporary* (London: Routledge 1998); Gerald L. Bruns, *Heidegger's Estrangements: Language Truth and Poetry in the Later Writings* (New Haven CT: Yale University Press 1989); and Gerald L. Bruns, *Maurice Blanchot: The Refusal of Philosophy* (Baltimore MD: Johns Hopkins University Press 1997).
11 See Maurice Blanchot, *The Infinite Conversation* [1969] trans. Susan Hanson (Minneapolis MN: University of Minnesota Press 1993); and Maurice Blanchot, *The Step (not) Beyond* [1973] trans. Lycette Nelson (Albany NY: SUNY Press 1992)

12 See Maurice Blanchot, *The Work of Fire* [1949] trans. Charlotte Mandell (Stanford CA: Stanford University Press 1995) pp. 300–44.

13 See Jacques Derrida, 'Living On – Border Lines' in Harold Bloom *et al. Deconstruction and Criticism* (New York: The Continuum Publishing Company 1979) pp. 75–176.

14 Jacques Derrida, *Aporias: Dying – Awaiting (One Another at) the 'Limits of Truth'* [1993] trans. Thomas Dutoit (Stanford CA: Stanford University Press 1993).

15 See Jacques Derrida, *Of Spirit: Heidegger and the Question* [1987] trans. Geoffrey Bennington and Rachel Bowlby (Chicago: University of Chicago Press 1991).

16 See Jacques Derrida, *Writing and Difference* [1967] trans. Alan Bass (London: Routledge 1978) pp. 79–153.

17 See Emmanuel Levinas, *Of God Who Comes to Mind* [1986] trans. Bettina Bergo (Stanford CA: Stanford University Press 1998).

18 See Martin Heidegger, *Being and Time*, division 2, ch. 2, section 57, 'The call of conscience'.

19 See the introduction by Gil Anidjar to Jacques Derrida, *Acts of Religion* (London: Routledge 2003) especially pp. 9–11.

20 See Emmanuel Levinas, *Totality and Infinity: An Essay on Exteriority* [1961] trans. Alphonso Lingis (Pittsburgh PA: Duquesne University Press 1969).

21 See Jacques Derrida, *Specters of Marx: The State of the Debt, the Work of Mourning and the New International* [1994] trans. Peggy Kamuf (London: Routledge 1994), on Blanchot esp. pp. 16, 29–31, and on Husserl see esp. ch. 5, 'Apparition of the inapparent: the phenomenological 'conjuring trick'", esp. footnote 5, p. 189, concerning the irreality of the noema. Derrida appeals to Husserl's categories when he writes of Marx's analyses of Stirner,

> This resembles an *epokhe*, a phenomenological reduction *of the* ghost, but Marx criticises it as a phenomenological reduction *to the* ghost (to the phenomenality or phantasm of a phantom). The reduction as subjectivization of the corporeal form of the external phantom is but a super-idealization and a supplementary spectralization.
>
> (pp. 129–30)

22 See Edmund Husserl, *Phantasy, Image Consciousness and Memory (1898–1925)* (HUA XXIII) trans. John Barnet Brough (Dordrecht: Springer 2005).

23 See Franz Rosenzweig, *The Star of Redemption* [1921] trans. William W. Hallo (Boston MA: Beacon Press 1971). Levinas invokes this text in the preface to *Totality and Infinity: An Essay on Exteriority* (1961). For discussion of the connection between them see Robert Gibbs, *Correlations in Rosenzweig and Levinas* (Princeton NJ: Princeton University Press 1992)

24 Maurice Blanchot, 'The Book to Come', trans. Sacha Rabinovitch, in *Maurice Blanchot: The Siren's Song*, ed. Gabriel Josipovici (Brighton: Harvester 1982) pp. 227–448.

25 On originary mourning, see the interview in *Points ... Interviews 1974–1994*, ed. Elisabeth Weber, trans. Peggy Kamuf *et al.* (Stanford CA: Stanford University Press 1995), '*Istrice 2: Ick buenn all hier*', pp. 300–26, in which Derrida also averts to a certain political theology in Heidegger's writings; 'a political theology that is both negative and resigned, and a poetology that takes the place of the active aspirations of a thinking that in *Being and Time* had shown clear pragmatic traits' (p. 310). Derrida is there reported as saying

> Even before the death of the other, the inscription in me of her or his mortality constitutes me. I mourn therefore I am, I am – dead with the death of the other, my relation to myself is first of all plunged into mourning, a mourning that is moreover impossible. This is also what I call ex-appropriation, appropriation caught in a double bind: I must and I must not take the other into myself;

> mourning is an unfaithful fidelity if it succeeds in interiorizing the other ideally in me, that is, in not respecting his or her infinite exteriority.
>
> (p. 321)

For a defining account of the importance for Derrida of mourning, see Pascale Anne Brault and Michael Naas, introduction to Jacques Derrida, *The Work of Mourning,* edited, translated and introduced by Pascale Anne Brault and Michael Naas (Stanford CA: Stanford University Press 2002).

26 See Edmund Husserl, *Zur Phaenomenologie des Intersubjektivitaet 1929–1935*, HUA 15, ed. Iso Kern (The Hague: Martinus Nijhoff 1972), p. 670, cited and trans. Dermot Moran, *Edmund Husserl: Founder of Phenomenology* (Oxford: Polity Press 2005):

> The absolute is nothing other than absolute temporalization and even its interpretation as the absolute which I directly encounter as my stationary streaming primordiality is a temporalization, a temporalization of this into something primarily existing. Therefore the absolute totality of monads, i.e. the primordiality of all the monads only exists by virtue of its temporalization.
>
> (p. 242)

27 See Jacques Derrida, *Psyche: Inventions of the Other,* trans. Catherine Porter, in Lindsay Waters and Wlad Godzich (eds) *Reading de Man Reading* (Minneapolis MN: University of Minnesota Press 1989) pp. 25–65.

28 The moments at which confession becomes the focus for concern are in the readings of Rousseau, in *Of Grammatology* (1967), in the readings of Augustine, suggested in 'How to Avoid Speaking: Denials', in *Memoires of the Blind* (1990), and of course in *'Circonfession'*, coming emphatically to the fore in *L'animal que donc je suis* [2006] op. cit.

29 For such a splitting in Kant, see Derrida's essay 'The University without Condition', in which the various different deployments of an 'as if' for Kant are set out, suggesting various different lines of entry for Kantian critique into an analysis of futurity. See Jacques Derrida, 'The Future of the Profession or the University without Condition', in Tom Cohen (ed.) *Jacques Derrida and the Humanities: A Critical Reader* (Cambridge: Cambridge University Press 2001) pp. 24–57; and in Jacques Derrida, *Without Alibi,* trans. Peggy Kamuf (Stanford CA: Stanford University Press 2002) pp. 202–37.

30 See Jacques Derrida, 'At the very moment in this work here I am' [1980], in Robert Bernasconi and Simon Critchley (eds) *Re-reading Levinas* (London: Athlone 1991) pp. 11–65, and for an analysis of these moves see Simon Critchley, *The Ethics of Deconstruction: Derrida and Levinas* (Oxford: Blackwell 1994).

31 For discussions of this, see Theodore Kisiel, *The Genesis of Being and Time* (Berkeley CA: University of California Press 1993). See also John van Buren, *Heidegger: Rumor of a Hidden King* (Bloomington IN: Indiana University Press 1995).

32 See Martin Heidegger, *Nietzsche,* 2 vols [1961] ed. and trans. David Farrell Krell, 4 vols (New York: Harper and Row 1970–).

33 See Maurice Blanchot, *The Infinite Conversation* [1969] trans. Susan Hanson (1993), on the relation at stake in research:

> This relation is such that it involves the absence of a common measure, the absence of a common denominator, and thus in some sense, the absence of relation between the terms: an exorbitant relation. Hence, the concern for marking either the interruption and rupture or the density and the plenitude of this field resulting from difference and tension. One can see as well, however, that continuity risks being merely the continuity of simple development, thus eliminating the irregularity of the 'curvature' or discontinuity risks being the simple juxtaposition of indifferent terms.
>
> (p. 7)

34 See Maurice Blanchot, 'Our Clandestine Companion', in Richard Cohen (ed.) *Face to Face with Levinas* (Albany NY: SUNY Press 1985).
35 See Emmanuel Levinas, 'Wholly Otherwise', pp. 3–10, in Bernasconi and Critchley (eds) *Re-reading Levinas* (op. cit.).
36 See Jacques Derrida, 'Hostipitality', in Derrida, *Acts of Religion*, ed. Gil Anidjar (London and New York: Routledge 2002) pp. 356–420, esp. pp. 394–95, and Emmanuel Levinas, *Existents and Existence* [1948] trans. Alphonso Lingis (op. cit.).
37 See Edmund Husserl, *Ideas One* (HUA 3, 1–2) section 49: 'Absolute Consciousness as the residuum of the abolition of the world'.
38 See Philippe Lacoue-Labarthe and Jean-Luc Nancy, *The Literary Absolute: Theory of Literature in German Romanticism* [1972] trans. Philip Barnard and Cheryl Lester (Albany NY: SUNY Press 1988).
39 Martin Heidegger, *On the Way to Language* [1959] trans. P. D. Hertz (New York: Harper and Row 1972).
40 For the obscure remark 'Only a new God can save us now', see Martin Heidegger, 'The Spiegel Interview' [1966] in Gunther Neske and Emil Kettering (eds) *Martin Heidegger and National Socialism: Questions and Answers* [1988] trans. Lisa Harries (New York: Paragon House 1990) pp. 41–66.
41 See Jacques Derrida, '*Geschlecht* II: Heidegger's Hand', trans. John P. Leavey Jr, in John Sallis (ed.) *Deconstruction and Philosophy: The Texts of Jacques Derrida* (Chicago: University of Chicago Press 1987) pp. 161–96.
42 See Jacques Derrida, '*Ousia* and *gramme*: Note on a Note in *Being and Time*', trans. Alan Bass, in Jacques Derrida, *Margins of Philosophy* (Brighton, UK: Harvester 1983) pp. 29–68.
43 For 'Anaximander's Saying', see Martin Heidegger, *Off the Beaten Track* [1950] trans. Kenneth Haynes and Julian Young (Cambridge: Cambridge University Press 2000).
44 See Emmanuel Levinas, 'The Trace of the Other', trans. Alphonso Lingis, in Mark C. Taylor (ed.) *Deconstruction in Context: Literature and Philosophy* (Chicago: University of Chicago Press 1982) pp. 345–59.

3.2 Literature, in the place of religion

1 See Maurice Blanchot, *The Work of Fire* [1949] trans. Charlotte Mandell (Stanford CA: Stanford University Press 1995).
2 See Jacques Derrida, '*Maurice Blanchot est mort*', in Christophe Bident and Pierre Vilar (eds) *Maurice Blanchot: récits critiques* (Paris: Editions Farrago/Editions Leo Scheer 2005) pp. 595–637.
3 The reason presented in *Groundwork for a Metaphysics of Morals* [1785] trans. H. J. Paton in *The Moral Law* (London: Hutchinson University Library 1948) is of course that to will one's own death is not universalisable (AK 6. 422). This duty to preserve one's life is the rational form of the commitments of faith.

> When on the contrary, disappointment and hopeless misery have quite taken away the taste for life; when a wretched man, strong in soul and more angered at his fate than faint hearted or cast down, longs for death and still preserves life without loving it – not from inclination or fear but from duty; then indeed his maxim has a moral content.
>
> (AK VI, 398)

4 See also Giorgio Agamben in *Homer Sacer: Sovereign Power and Bare Life* [1994] trans. Daniel Heller Roazen (Stanford CA: Stanford University Press 1998).
5 The term 'existential' is formed on the same principle as 'quale' and 'qualia', with a singular 'existentiale', and a plural 'existentialia'. They are those determinations distinctive of Dasein.

6 Jacques Derrida, *Aporias: Dying – Awaiting (One Another at) the 'Limits of Truth'* [1993] trans. Thomas Dutoit (Stanford CA: Stanford University Press 1993).
7 See Maurice Blanchot, *The Writing of the Disaster* [1980] trans. Ann Smock (Lincoln NE and London: University of Nebraska Press 1986).
8 For discussions of this, see the papers gathered together in Jeffrey Librett (ed.) *Of the Sublime: Presence in Question* [1988] trans. Jeffrey Librett (Albany NY: SUNY Press 1993). See also the essay by Jeffrey S. Librett, 'Positing the Sublime: Reading Heidegger Reading Kant', for the Kant of Heidegger's 1962 essay, 'Kant's Thesis on Being', ibid., pp. 193–219.
9 See Emmanuel Levinas, *God, Death and Time* [1976] trans. Bettina Bergo (Stanford CA: Stanford University Press 2000) esp. part I, 'Death and time'.
10 See Martin Heidegger, *The Metaphysical Foundations of Logic Beginning with Leibniz* (GA 26, 1978) trans. Michael Heim (Bloomington IN: Indiana University Press 1984).
11 See Jacques Derrida, '*Geschlecht I*: Sexual Difference, Ontological Difference', [1983] in Peggy Kamuf (ed.) *Jacques Derrida: Between the Blinds* (Brighton, UK: Harvester 1994).
12 See Martin Heidegger, *Off the Beaten Track* [1950], ed. and trans. Julian Young and Kenneth Haynes (Cambridge: Cambridge University Press 2002).
13 See Martin Heidegger, 'The Spiegel Interview' [1966] in Gunther Neske and Emil Kettering (eds) *Martin Heidegger and National Socialism: Questions and Answers* [1988] trans. Lisa Harries (New York: Paragon House 1990) pp. 41–66.
14 For a discussion drawing attention to this shift see Levinas, *God, Death and Time*, op.cit: 'The Radical Question: Kant against Heidegger', pp. 57–61; and 'A Reading of Kant Continued' (L: GDT, pp. 62–65). 'The practical philosophy of Kant shows that the Heideggerian reduction is not obligatory. It shows that there might be in the history of philosophy a signification other than that of finitude' (p. 61). I shall pursue this in Part V of these enquiries. It opens out a possible distinction between a phenomenological reduction of death to an existential structure, and a further eidetic reduction of that existential structure to a finitude of time. A link to Blanchot's thinking of an elongation of time emerges:

> This orientation does not refute being-towards death, which is according to Heidegger the presupposition of finitude. There lies the great force of Kant's practical philosophy: the possibility of thinking a beyond of time by way of hope, but obviously not a beyond that would prolong time, not a beyond that is (and would be). But neither an everyday derivative of originary time.
>
> (L: GDT, p. 63)

15 See Emmanuel Levinas, 'Ethics as First Philosophy' [1984] trans. Sean Hand and Michael Temple, in Sean Hand (ed.) *The Levinas Reader* (Oxford: Blackwell 1989) pp. 75–87.
16 See Emmanuel Levinas, 'Intentionality and Sensation', in Emmanuel Levinas, *Discovering Existence with Husserl*, trans. Richard A. Cohen and Michael B. Smith (Evanston IL: Northwestern University Press 1998) pp. 135–51.

4.1 Phenomenology as democracy to come

1 There are two versions of the dialogue from which this extract is taken. The first was contributed as an appendix to the discussion in Harold Coward and Toby Foshay (eds) *Derrida and Negative Theology* (Albany NY: SUNY Press 1992) under the title '*Post-scriptum: Aporias*, Ways and Voices', trans. John P. Leavey, pp. 283–323; and the second as one of the three texts translated into English under the collective title *On the Name*, ed. Thomas Dutoit (Stanford CA: Stanford University Press 1995), '*Sauf le nom: post-scriptum*', pp. 35–85.

2 The connection from *différance* to the thinking of negative theology is marked and refused already in the essay '*Différance*' (Jacques Derrida, *Speech and Phenomena and Other Essays*, trans. David Allison, Evanston IL: Northwestern University Press 1973, pp. 129–60:

> And we will have to point out everything that it is not and consequently that it has neither existence nor essence. It belongs to no category of being, present or absent. And yet what is thus denoted as difference is not theological, not even in the most negative order of negative theology. The latter, as we know, is always occupied with letting a supraessential relation go beyond the finite categories of essence and existence, that is, of presence, and always hastens to remind us that, if we deny the predicate of existence to God, it is in order to recognize him as superior, in conceivable and ineffable mode of being.
>
> (D: SP 1967, 134)

3 See Augustine, *Confessions*, trans. Henry Chadwick (Oxford: Oxford University Press 1991).
4 See Jacques Derrida, *Politics of Friendship* [1994] trans. George Collins (London: Verso 1997) which according to Derrida's 'Foreword' derives from a seminar given in 1988–89.
5 Jacques Derrida, *The Other Heading: Reflections on Today's Europe* [1991] trans. Pascale Anne Brault and Michael Naas (Bloomington IN: Indiana University Press 1992); and *The Gift of Death* [1992] trans. David Wills (Chicago: University of Chicago Press 1995), especially the discussion of Jan Patoscka in the first two chapters.
6 The main lines are the series of writings on negative theology, the line of which is to be in part reconstructed here, juxtaposed with the more obviously political texts: *The Other Heading: Reflections on Today's Europe* [1991] op. cit.; *Specters of Marx: The State of the Debt, The Work of Mourning and the New International* [1993] trans. Peggy Kamuf (London and New York: Routledge 1994); *Politics of Friendship* [1994] op. cit., and *Voyous: Rogues: Two Essays on Reason* [2003] trans. Pascale Anne Brault and Michael B. Naas (Stanford CA: Stanford University Press 2005). *Specters* proved highly controversial, prompting the publication of several discussion volumes, amongst others Michael Sprinker (ed.) *Ghostly Demarcations: A Symposium on Jacques Derrida's 'Specters of Marx'* (London: Verso 1999), which includes Jacques Derrida's essay 'Marx and Sons', pp. 213–69. There he replies to some of his critics and, prompted by Antonio Negri's essay 'The Smile of the Specter', announces a certain clandestine identification of Spinoza, Marx and himself as Marrano, the extreme case of which is made up of those who do not know themselves to be Marrano. The notion of the Marrano is also invoked in conclusion to *Aporias*, and will be elucidated in the exposition of the *topolitology* of the secret, in Part IV, Section 2.
7 See S. S. Prawer, *Karl Marx and World Literature* (Oxford: Oxford University Press 1978).
8 This analysis is borrowed from Len Lawlor, in the final chapter of his study, *Derrida and Husserl: The Basic Problem of Phenomenology* (Bloomington IN: Indiana University Press 2002), in which he marks the shift from the analysis of genesis and of the sign to these analyses of the promise, of mourning and of hospitality. The promise opens the future out for the arrival of alterity and indeed of time.
9 For a discussion of this and for an analysis of the emergence of Derrida's account of politics out of the readings of Hegel and Kant, of Levinas and Heidegger, see Richard Beardsworth, *Derrida and the Political* (London: Routledge 1996). See also the essay by Geoffrey Bennington, 'Derrida and Politics', in Tom Cohen (ed.) *Jacques Derrida and the Humanities: A Critical Reader* (Cambridge: Cambridge University Press 2001) pp. 193–212.
10 See Bernard Stiegler, *Technics and Time One: The Fault of Epimetheus* [1994] trans. Richard Beardsworth and George Collins (Stanford CA: Stanford University Press 1998). Volume 2 has yet to appear in translation. Stiegler subjects Heidegger's thinking of time and temporality to a thorough interrogation under the guidance of the

suspicion that technicity, or technique, a framing of what there is in accordance with an inherited set of capacities and preoccupations, already plays a role in the articulation of time in *Being and Time.* Thus the distinction between the temporality of Dasein and that of readiness-to-hand is disrupted. Stiegler also identifies the proximity between the aporetics of deconstruction and the *aporia* of Husserl's passive synthesis: see his essay 'Derrida and Technology: Fidelity at the Limits of Deconstruction and the Prosthesis of Faith', in Cohen (ed.) *Jacques Derrida and the Humanities* (op. cit.) pp. 238–70.

11 For a discussion of Heidegger on this complex relation see Joanna Hodge, 'Freedom, *Phusis*, *Techne*: Thinking the Inhuman', in John Joughin (ed.) *Reflections on the Inhuman* (Manchester: Manchester University Press 2000).

12 See Francis Fukuyama, *The End of History and the Last Man* (New York: Free Press 1992).

13 See Karl Marx and Friedrich Engels, *The German Ideology* [1846] in *Marx and Engels Collected Works* (New York: International Publishers 1976).

14 See Maurice Blanchot, 'Marx's Three Voices', in Blanchot, *Friendship* [1971] trans. Elizabeth Rottenberg (Stanford CA: Stanford University Press 1997) pp. 98–100. The three voices are described by Blanchot as producing writings sufficiently distinct as to defy translation from one into the others.

15 Martin Heidegger, 'Anaximander Fragment' [1946] in Heidegger, *Off the Beaten Track* [1950] trans. Julian Young (Cambridge: Cambridge University Press 2001).

16 Edmund Husserl, 'Philosophy as Rigorous Science' [1911] in Peter McCormick and Frederick Elliston (eds) *Husserl: Shorter Works* (Notre Dame IN: University of Notre Dame Press 1981) pp. 159–98; and Edmund Husserl, *The Crisis of the European Sciences and Transcendental Phenomenology* [1939] trans. David Carr (Evanston IL: Northwestern University Press 1973).

17 See Jacques Derrida, 'From Restricted to General Economy: A Hegelianism without Reserve', in Derrida, *Writing and Difference* [1967] trans. Alan Bass (London and New York: Routledge 1978) pp. 251–77). For Georges Bataille see *Inner Experience* [1943, 1953] trans. Leslie Anne Boldt (Albany NY: SUNY Press 1988). The notion of general economy expands the scope of political economy to include all organisation and dispensing of energy.

18 See Jean Luc Marion, *God without Being* [1983] trans. Thomas A. Carlson (Chicago IL: University of Chicago Press 1991).

19 On this see Peter Osborne, 'Small Scale Victories, Large Scale Ddefeats: Walter Benjamin's Politics of Time', in Andrew Benjamin and Peter Osborne (eds) *Walter Benjamin's Philosophy: Destruction and Experience I*, 2nd edn (Manchester: Clinamen Press 2000) pp. 56–98, and on the 'secret cargo' of surrealism, see p. 92.

20 See the editor's note to Jacques Derrida, *On the Name* [1993] edited by Thomas Dutoit, from translations by Geoffrey Bennington, John P. Leavey Jr and David Wood (Stanford CA: Stanford University Press 1993). Of the three essays in *On the Name*, 'Passions', translated by David Wood, appeared originally in English and in a shorter form in Derrida, *A Critical Reader*, ed. David Wood (Oxford: Blackwell 1992). An earlier shorter version of *Sauf le nom* appeared for the first time also in English in the translation of John P. Leavey Jr, under the title 'Post-scriptum' in a volume edited by Harold Coward and Toby Foshay (Albany NY: SUNY Press 1992). They appear here by courtesy of these publishers. *Khora*, translated by Geoffrey Bennington, appears here for the first time in English, in a revised version from how it originally appears in French in *Poikilia: études offertes à Jean-Pierre Vernant* (Paris: Ecole des Hautes Etudes en Sciences Sociales 1987).

21 See Arthur Bradley, *Negative Theology and Modern French Philosophy* (London: Routledge 2004), which takes a reading of *Sauf le Nom* for its frame, while also discussing aspects of the work of Jean-Luc Marion, Michel de Certeau, Michel Foucault and Julia Kristeva. See also Hent de Vries, *Philosophy and the Turn to Religion* (Baltimore MD: Johns Hopkins University Press 1999); and Hent de Vries, *Religion and Violence: Philo-*

sophical Perspectives from Kant to Derrida (Baltimore MD: Johns Hopkins University Press 2002). He takes the trajectory of Derrida's writings from the early 1980s onwards as performing not so much an ethical turn as a religious turn. All three provide excellent and indispensable readings.

While working from within the tradition which privileges a Christian hegemony over other religions, Derrida is nevertheless more aware of the restrictions and distortions thereby imposed. There are remarks to this effect in 'Faith and Knowledge: The Two sources of "Religion" at the Limits of Reason Alone', which engages with the privilege to Christianity installed in the writings of Kant, of Hegel and of Bergson, and indeed states, in italics, 'To think "religion" is to think the "Roman"' (*Acts of Religion*, p. 45). Gil Anidjar in his introduction to Jacques Derrida, *Acts of Religion* (London: Routledge 2002) draws attention to Derrida's rewriting of *Donner la mort* (1992, 1999) for publication in 1999, to include a consideration of the elision of the fate of Ishmael from accounts of the sacrifice of Isaac. This points up the question to political theologies of what happens to the peoples who are not chosen. Neither Bradley nor Hent de Vries read Derrida's engagements with negative theology as opening onto a critical political theology, based on the thought of there being as many Gods as there are names for God, as I propose to do, and as critically in dispute with Benjamin's rewriting of theology. The key differences here concern the temporal conditions under which divinity is understood to enter into human time. For the discussion of the writings of Benjamin in relation to those of Derrida, various essays by Werner Hamacher have proved of inestimable assistance.

22 The trajectory of this thinking thus can be seen to start variously with the remarks about negative theology in the essay '*Différance*' (1968), or with the miming of apocalyptic writing in 'Of an Apocalyptic Tone Newly Adopted in Philosophy' (1980). In the latter Derrida reaffirms the doubts first brought to the fore in 'Violence and Metaphysics: An Essay on the Thought of Emmanuel Levinas' (1963), about both the eschatology of finitude, proposed by Heidegger, and the eschatology proposed by Levinas, providing him with an exit from their respective reconfigurations of Husserl's phenomenology. For my reading of 'Violence and Metaphysics', I am indebted to John Llewelyn who, a very long time ago, taught a course I was fortunate to attend, in which he opened out the text for discussion in an exemplary fashion.

4.2 'There is the secret'

1 See Werner Hamacher, 'Cult History: Benjamin's Sketch "Capitalism as Religion"', *Diacritics: Journal of Contemporary Criticism*, vol. 32, nos. 3–4, pp. 81–106.

2 For an earlier version of this argument, with which the following is not entirely compatible, see Joanna Hodge, 'The Timing of Elective Affinity: Walter Benjamin's Strong Aesthetics', in Andrew Benjamin (ed.) *Walter Benjamin and Art* (London: Continuum 2005) pp. 14–30.

3 See Eric Alliez, *Capital Times: Tales from the Conquest of Time* [1991] trans. Georges van den Abbeele (Minneapolis MN and London: University of Minnesota Press 1996) who quotes Heidegger, 'The Concept of Experience', from Martin Heidegger, *Off the Beaten Track* [1950] trans. Julian Young and Kenneth Haynes (Cambridge: Cambridge University Press 2002) pp. 86–156:

> The science of absolute knowledge is the secular theology of the world not because it secularizes Christian and ecclesiastical theology but because it belongs in the nature of ontology. Ontology is older than any Christian theology, and Christian theology in turn must first be realized before a process of secularization can lay hold of it. The theology of the Absolute is that knowledge of beings qua beings which among the Greek thinkers reveals and follows its onto-theological nature without ever pursuing it to its foundation. The language of absolute

science shows that Christian theology in what it knows and in the way it knows its knowledge is metaphysics.

(p. 152 trans. mod.)

4 For an account of this relation, focused on Levinas, Husserl and Kant, rather than on Heidegger, Derrida and Benjamin, see Joanna Hodge, 'Ethics and Time: Levinas between Kant and Husserl', *Diacritics: Journal for Contemporary Criticism*, vol. 32, nos. 3–4, pp. 107–35.
5 For Werner Hamacher's outstanding reading of the latter essay, see '"Now": Walter Benjamin on Historical Time', in Heidrun Friese (ed.) *The Moment: Time and the Rupture in Modern Thought* (Liverpool, UK: Liverpool University Press 2001) pp. 161–96. See p. 169, and: 'There is historical time only insofar as there is an excess of the unactualised, the unfinished, failed, thwarted, which leaps beyond its particular Now and demands from another Now its settlement, correction and fulfilment' (p. 164). He writes here of a 'theology of missed or distorted possibilities', of which it seems to me the encounter between Derrida and Benjamin is undoubtedly one.
6 This is developed in the analysis of the shipwreck of reason, set out as ambiguous between a deliberate and an accidental running aground, in the second of the two essays in Jacques Derrida, *Rogues: Two Essays on Reason* [2003] trans. Pascale Brault and Michael Naas (Stanford CA: Stanford University Press 2005) pp. 117–59.
7 See Jacques Derrida, *Acts of Religion*, ed. and trans. Gil Anidjar (London and New York: Routledge 2002) pp. 40–101.
8 Immanuel Kant, *Religion within the Boundaries of Mere Reason and Other Writings*, ed. Allen Wood and George di Giovanni, Cambridge Texts in the History of Philosophy (Cambridge: Cambridge University Press 1998).
9 See Jacques Derrida, *Truth in Painting* [1978] trans. Geoffrey Bennington and Ian McLeod (Chicago: Chicago University Press 1987) part I, 'Parergon', pp. 15–147, the second section of which is also called 'The parergon'. There Derrida writes: 'Kantian question: the relation of the concept to the non-concept (up/down, left/right), to the body, to the signature which is placed "on" the frame; in fact, sometimes, structurally, always. The prosthesis' (p. 78). See also p. 56 for the treatment of the four temptations of religious enthusiasm, in *Religion within the Limits of Reason Alone.*
10 See Jean-Luc Marion, *God without Being* [1983] trans. Thomas Carlson (Chicago: Chicago University Press 1991). For discussion of this and of the ongoing debate between Derrida and Marion, see John D. Caputo and Michael Scanlon (eds) *God, the Gift and Postmodernism* (Bloomington IN: Indiana University Press 1999); Robyn Horner, *Rethinking God as Gift: Marion, Derrida and the Limits of Phenomenology* (New York: Fordham University Press 2001); and Ian Leask and Eoin Cassidy (eds) *Givenness and God: Questions of Jean-Luc Marion* (New York: Fordham University Press 2005).
11 See Jean-Luc Nancy, 'Of Divine Places' [1984] trans. Michael Holland, to be found in *Paragraph: A Journal of the Modern Critical Theory Group*, vol. 7, 1986; and in Jean-Luc Nancy, *The Inoperative Community*, ed. Peter O'Connor (Minneapolis MN: University of Minnesota Press 1991) pp. 110–50.
12 See the papers collected together in Jeffrey Librett (ed.) *Of the Sublime: Presence in Question* [1988] (Albany NY: SUNY Press 1993).
13 For Derrida's poignant reading of Celan, 'Schibboleth', delivered 14 October 1984, see in Aris Fioretis (ed.) *Word Traces: Readings of Paul Celan* (Baltimore MD: Johns Hopkins University Press 1994).
14 See Jacques Derrida, *Of Spirit: Heidegger and the Question* [1987] trans. Geoffrey Bennington and Rachel Bowlby (Chicago IL: University of Chicago Press 1991).
15 There are of course many other allusions to Benjamin in Derrida's writings, amongst others in *Truth in Painting* (1978) (see above), a footnote in *Memoires: For Paul de Man* [1986] trans. Cecile Lindsay *et al.* (New York: Columbia University Press 1986, 1989) p. 153, and rather more in *Politics of Friendship* (1994). See also Hent de Vries, *Religion and Violence: Philosophical Perspectives from Kant to Derrida* (2002) especially chapter 3.

16 See Martin Heidegger, *Introduction to Metaphysics* [1953] trans. Gregory Fried and Richard Polt (New Haven CT: Yale University Press 2000) pp. 169–78.

17 Walter Benjamin, *Walter Benjamin: Selected Writings Volume Two: 1927–1934*, eds Michael W. Jennings, Howard Eiland and Gary Smith (Cambridge MA: Belknap Press of Harvard University Press 1999) pp. 541–42.

18 Even more telling is the following, again to Scholem, from 1 December 1920 in *Walter Benjamin Briefe*, eds Gershom Scholem and Theodor Adorno, Frankfurt am Main: Suhrkamp 1978:

> I have read Heidegger's book on Duns Scotus. It is unbelievable that someone can be habilitated with such a work, in the formation of which nothing more than assiduity and a mastery of scholastic Latin is necessary, and which, despite all its philosophical pretensions, is essentially only a piece of good translation.
>
> (p. 246)

See also David S. Ferris: 'Introduction: Aura, Resistance, Event of History', in David S. Ferris (ed.) *Walter Benjamin: Theoretical Questions* (Stanford CA: Stanford University Press 1994) pp. 1–26, and esp. p. 23, on the relation between Benjamin and Heidegger, and on Benjamin's relation to a phenomenology of discontinuity.

19 For a discussion of Benjamin and Heidegger on technology, in which Benjamin's distinctions between a first and a second technology, as drawn in 'The Work of Art in the Age of Its Mechanical Reproducibility' are connected to his notions of a first and second nature, see Beatrice Hanssen, 'Benjamin or Heidegger: Aesthetics and Politics in an Age of Technology', in Andrew Benjamin (ed.) *Walter Benjamin on Art* (London: Continuum 2005) pp. 73–90. See also Beatrice Hanssen's discussion of Benjamin on nature, history and natural history, in *Walter Benjamin's Other History: Of Stones, Animals, Human Being and Angels* (Berkeley CA: University of California Press 1998).

20 See Jacques Derrida, '*Des Tours de Babel*', trans. Joseph F. Graham, in Joseph F. Graham (ed.) *Difference in Translation* (Ithaca NY: Cornell University Press 1985) pp. 165–248.

21 See Jacques Derrida, 'Living On – Border Lines' (1978) in Harold Bloom *et al.* *Deconstruction and Criticism* (New York: The Continuum Publishing Company 1979).

22 This would bring Benjamin into an obscure proximity to the thinking of Husserl on pure grammar, a surmise which cannot here be followed up.

23 See Walter Benjamin, *Walter Benjamin: Selected Writings Volume One, 1913*–1926, eds Marcus Bullock and Michael W. Jennings (Cambridge MA: Belknap Press of Harvard University Press 1996) pp. 62–74 and pp. 253–64.

24 In his essay 'Conclusions: Walter Benjamin's "The Task of the Translator"', written up from notes from the sixth Messenger Lecture in March 1983, de Man notes the comical destiny of mistranslation suffered by this very essay, in both Maurice de Gandillac's French version and in Harry Zohn's English version, in Walter Benjamin, *Illuminations*, ed. Hannah Arendt (New York: Schocken Books 1969) pp. 69–82. See Paul de Man, *The Resistance to Theory* (Minneapolis MN: University of Minnesota Press 1987) pp. 73–93. In the discussion de Man intimates his notion of the inhuman potential of language, marked up by Neil Hertz and put in question by Meyer Abrams, see pp. 94–104.

25 See Paul de Man, *Aesthetic Ideology*, ed. Andrej Warminiski (Minneapolis MN: University of Minnesota Press 1996).

26 For some of the exchanges in which this disagreement found expression, see Ernst Bloch, Georg Lukács, Bertolt Brecht, Walter Benjamin and Theodor Adorno, *Aesthetics and Politics*, ed. Rodney Taylor (London: NLB/Verso 1977).

27 For the discussion of 9/11 see Giovanna Borradori (ed.) *Philosophy in a Time of Terror: Dialogues with Jürgen Habermas and Jacques Derrida* (Chicago: University of Chicago Press 2003).

5.1 Cultivating limits

1 Immanuel Kant, *Critique of Judgment* [1790] trans. Werner Pluhar (Indianapolis IN: Hackett 1987).

2 For Derrida, see Jacques Derrida, *Of Grammatology* [1967] trans. Gayatri Chakravorty Spivak (Baltimore MD: Johns Hopkins University Press 1978) on the supplement of the origin; and Jacques Derrida, *The Monolingualism of the Other: Prosthesis of the Origin* [1996] trans. Patrick Mensah (Stanford CA: Stanford University Press 1998). For Stiegler, see Bernard Stiegler, *Technics and Time 1: The Fault of Epimetheus* [1994] trans. Richard Beardsworth and George Collins (Stanford CA: Stanford University Press 1998).

3 For Aristotle, see Martha Craven Nussbaum, *Aristotle's De Motu Animalium: Text, Translation and Commentary* (Princeton NJ: Princeton University Press 1978).

4 For Heidegger, see Martin Heidegger, 'Phenomenological Interpretation in Connection with Aristotle: An Indication of a Hermeneutical Situation' [1922] in John van Buren (ed.) *Martin Heidegger: Supplement from the Earliest Essays to 'Being and Time' and Beyond* (Albany NY: SUNY Press 2002) pp. 111–46; Martin Heidegger, *Phenomenological Interpretation of Aristotle: Introduction to Philosophical Research (1921–1922)*, GA 61, trans. Richard Rojcewicz (Bloomington IN: Indiana University Press 2001); and Martin Heidegger, *Aristotle's Metaphysics Theta 1–3, on the Essence and Actuality of Force* [1931] GA 33, trans. Walter Brogan and Pater Warnek (Bloomington IN: Indiana University Press 1995).

5 It would be necessary to set this out as a contrast to the notion of machines, as developed by Deleuze and Guattari, to develop this alternative to the supposed exclusive disjunction of natural causality and human intending. The notion of originary contamination as prosthetic similarly takes up the challenge of transforming the terms of reference assumed set down by a shared inheritance, the transmission of which each is attempting to transform.

6 See Jacques Derrida, 'The Animal Which Therefore I Am (More to Follow)', trans. David Willis, in *Critical Inquiry* (Chicago: University of Chicago Press) vol. 28, no. 2, winter 2002, pp. 369–418, cited as D: 1997, AA; and Jacques Derrida, 'Typewriter Ribbon: Limited Ink (2)', trans. Peggy Kamuf, in Jacques Derrida, *Without Alibi* (Stanford CA: Stanford University Press 2002) pp. 71–160, to be cited as D: 1998, WA. The latter appeared first in a slightly different version in Tom Cohen (ed.) *Material Events: Paul de Man and the Afterlife of Theory* (Minneapolis MN: University of Minnesota Press 2001). In the course of it, Derrida remarks:

> I will add two subtitles to my title, namely 'machine' and 'textual event'. These are words de Man uses in 'Excuses (*Confessions*)'. I will thus propose that we interrogate together, at least obliquely, the use of these words, 'machine' and 'textual event' in *Allegories of Reading*. Their use as well as their supposed meaning. My hypothesis is that de Man reinvents and signs these words in a certain way even as he leads us, if we can still put it that way, toward the 'thinking of materiality' that comes to light in *Aesthetic Ideology*. The coherent use, the performative inaugurality of these words ('machine' and 'textual event'), their conceptual effects and the formalization that will follow, in semantics and beyond semantics, this is what will affect in a necessary fashion all of de Man's writing.
>
> (D: 1998, WA p. 79)

The notions of machine and textual event are replayed to unsettle the distinctions argued for by Kant in the Third Critique between the organic and the inorganic, and between the final causes of human judgment, and causality as a mechanism of nature. For Paul de Man, see his *Allegories of Reading: Figural Language in Rousseau, Nietzsche, Rilke, and Proust* (New Haven CT and London: Yale University Press 1979).

7 See Jacques Derrida, 'The Animal Which Therefore I Am (More to Follow)' (1997) trans. David Willis. This might be better translated as 'Animal Therefore I Am: By

Descent' since it is a play on the famous phrase of Descartes: '*cogito ergo sum, je pense donc je suis*'. I think therefore I am. The implication is that the deduction would work better if rephrased, 'I am animal therefore I am.' See also Jacques Derrida, *L'animal que donc je suis*, édition établie par Marie-Louise Mallet (Paris: Editions Galilée 2006) to be cited as D: 2006, AQD, of which Wills translated pp. 15–77, and in which the remaining sections on Descartes and Kant, and on Levinas and Lacan are to be found.

8 See Immanuel Kant, *Anthropology from a Pragmatic Point of View* [1798] trans. Mary J. Gregor (The Hague: Martinus Nijhoff 1974).

9 This essay was delivered to a conference in Montreal in 1971, and is printed in Jacques Derrida, *Margins of Philosophy* [1972] trans. Alan Bass (Brighton, UK: Harvester 1982). The encounter with John Searle is marked up in the publication of Derrida's contributions to it in Jacques Derrida, *Limited Inc.* (Evanston IL: Northwestern University Press 1988, 1990) with its important afterword: 'Afterword: Toward an Ethic of Discussion', pp. 111–60. See also Jacques Derrida, *Memoires: For Paul de Man*, 2 edns (New York and London: Columbia University Press 1986, 1989).

10 Derrida's accounts of time and of history are critically opposed to that developed by Paul Ricoeur in *Time and Narrative* [1983] 3 vols, trans. Kathleen McLaughlin and David Pellauer (Chicago: University of Chicago Press 1984). This develops and supplements the disagreement between Derrida and Ricoeur, on metaphor, as evidenced in Paul Ricoeur, *The Rule of Metaphor* [1975] trans. Robert Czerny (London: Routledge 1978); and Jacques Derrida, 'The *Retrait* of the Metaphor', *Enclitic 2*, no. 2 (1978) pp. 5–33.

11 See Jacques Derrida and Geoffrey Bennington, *Jacques Derrida* [1991] trans. Geoffrey Bennington (Chicago: University of Chicago Press 1993).

12 For a detailed response to the question of autobiography for Jacques Derrida, and for a response to his '*circonfession*', see Hélène Cixous, *Portrait of Jacques Derrida as a Young Jewish Saint* [2001] trans. Beverley Bie Brahic (New York: Columbia University Press 2004): '*Le monolinguisme de l'autre* tells us everything there is to know about the mental anguish to which we owe our books of memoirs' (p. 118). For the latter, see Jacques Derrida, *The Monolingualism of the Other* [1996] trans. Patrick Mensah (Stanford CA: Stanford University Press 1999).

13 See Jacques Derrida, 'Living On – Border Lines', in Harold Bloom, Paul de Man, Jacques Derrida, Geoffrey Hartman and J. Hillis Miller, *Deconstruction and Criticism* (New York: Continuum 1992) pp. 75–175, especially for the transition from *Übersetzung* to *Übertragung*, see pp. 87–88, and from *Übersetzung* to *Überwindung*, see pp. 102–6.

14 See Jacques Derrida, *The Ear of the Other: Otobiography, Transference, Translation*, English edn, ed. Christie McDonald (New York: Schocken Books 1985). See also the various papers collected in Diane P. Michelfelder and Richard E. Palmer (eds) *Dialogue and Deconstruction: The Gadamer/Derrida Encounter* (Albany NY: SUNY Press 1989).

15 See Jacques Derrida, '*Geschlecht*: Sexual Difference, Ontological Difference' [1983] in *A Derrida Reader: Between the Blinds*, ed. Peggy Kamuf (London and New York: Harvester Wheatsheaf 1991); Jacques Derrida, '*Geschlecht* II: Heidegger's Hand', trans. John P. Leavey Jr, in John Sallis (ed.) *Deconstruction and Philosophy: The Texts of Jacques Derrida* (Chicago: University of Chicago Press 1987) pp.161–96; and Jacques Derrida, 'Heidegger's Ear: *Philopolemology* (*Geschlecht* IV)' [1989] trans. John P. Leavey Jr, in John Sallis (ed.) *Reading Heidegger: Commemorations* (Bloomington IN: Indiana University Press 1993) pp. 163–220. The possibility that the thinking of *Geschlecht* marks Heidegger's encounter with the task of rewriting transcendental aesthetics cannot here be pursued.

16 See Paul de Man, *Aesthetic Ideology*, ed. Andrzej Warminski (Minneapolis MN: University of Minnesota Press 1996) pp. 70–90, which contains versions of the third, fourth and fifth of the Messenger Lectures delivered by de Man in March 1983. 'Phenomenality and Materiality in Kant' is a draft of the third lecture.

17 See Jacques Derrida, 'Structure, Sign and Play in the Human Sciences', in Derrida, *Writing and Difference* [1967] trans. Alan Bass (London: Routledge 1978) pp. 278–94.

See also de Man's essay, 'The Rhetoric of Blindness: Jacques Derrida's Reading of Rousseau', in *Blindness and Insight: Essays in the Rhetoric of Contemporary Criticism* (London: Methuen 1983) in which he diagnoses in *Of Grammatology* Derrida's redeployment of certain figural moves made by Rousseau. De Man writes:

> In France it took the rigor and intellectual integrity of a philosopher whose main concern is not with literary texts to restore the complexities of reading to the dignity of a philosophical question. . . .
>
> Jacques Derrida makes the movements of his own reading an integral part of a major statement about the nature of language in general. His knowledge stems from the actual encounter with texts, with a full awareness of the complexities involved in such an encounter.
>
> (pp. 110–11)

18 See *Memoires* (op. cit.) 2nd edn, pp. 156–261.
19 For a discussion of the various Kantianisms discussed by Derrida at various stages in his trajectory and in various texts, see the readings collected together in Philip Rothfield (ed.) *Kant after Derrida* (Manchester: Clinamen Press 2003). For a discussion of Derrida's writings as a response to the gap opened up in Kant's texts by Hegel's critiques of them, see Richard Beardsworth in *Derrida and the Political* (London: Routledge 1996) especially chapter 2.
20 See Jean-Luc Marion, *God without Being* [1983] trans. Thomas Carlson (Chicago: University of Chicago Press 1991), and for a response to this see Dominique Janicaud, *Phenomenology and the 'Theological Turn'* [1991] trans. Bernard G. Prusak (New York: Fordham University Press 2000).
21 See Friedrich Nietzsche, *On the Genealogy of Morals* [1887] trans. R. J. Hollingdale and W. Kaufmann, in Walter Kaufman (ed.) *On the Genealogy of Morals and Ecce Homo* (New York: Vintage Books 1969). For lying see 3.19, for the surmised right to make promises, and, implicitly on lies, see 2.1.
22 See Paul de Man, 'Conclusions: Walter Benjamin's "The Task of the Translator"', in Paul de Man, *The Resistance to Theory* (Minneapolis MN: University of Minnesota Press 1986) pp. 73–105, at 86–87. This was delivered as the conclusion to the above-mentioned series of Messenger Lectures.
23 These points of contact, tangents, and touches should of course be read in the context of Derrida, *On Touching: Jean-Luc Nancy* [2000] trans. Christine Irizarry (Stanford CA: Stanford University Press 2005). The manner in which the movement of *circonfession* displaces ellipsis as the characteristic gesture of reading made by Derrida should also be noted.
24 See *Walter Benjamin: Selected Writings*, 5 vols, eds Marcus Bullock and Michael Jennings (Cambridge MA: The Belknap Press of Harvard University Press 1996–) vol. 1 (1996).
25 Martin Heidegger, *Being and Time* [1927] trans. John Macquarrie and Edward Robinson (Oxford: Blackwell 1962).
26 See Jacques Derrida, 'Faith and Knowledge: The Two Sources of "Religion" at the Limits of Reason Alone' [1996] in Jacques Derrida, *Acts of Religion*, ed. Gil Anidjar (New York and London: Routledge 2002) pp. 42–101. See also the second essay in Jacques Derrida, *Rogues: Two Essays on Reason* [2003] trans. Pascale Anne Brault and Michael Naas (Stanford CA: Stanford University Press 2005) 'The "World" of the Enlightenment to Come: Exception, Calculation and Sovereignty', pp. 118–60.
27 See Immanuel Kant, 'Idea for a Universal History', in Immanuel Kant, *Political Writings*, ed. Hans Reis (Cambridge: Cambridge University Press 1998). See also Gottfried Hoeffe, *Kant's Cosmopolitan Theory of Law and* Peace [2001] trans. Alexandra Newton (Cambridge: Cambridge University Press 2006).

28 Thus briefly it seems to me to be an error to use a Kantian distinction between an idealism and materialism to produce a diagnosis of Heidegger as endorsing an idealism with respect to temporality. While temporality is an articulation of Dasein, Dasein is not simply an articulation of consciousness, but precisely its necessarily material instance. Dasein is given as a set of spatio-temporal coordinates with an historical index. The notion of idealism as deployed by Heidegger is thus to be distinguished from that deployed by Kant. See Wilhelm Blattner, *Heidegger's Temporal Idealism* (Cambridge: Cambridge University Press 1999).

29 Martin Heidegger, *The Fundamental Concepts of Metaphysics: World, Finitude, Solitude*, trans. Will McNeill and Nicholas Walker (Bloomington IN: Indiana University Press 1995) sections 45–63.

30 See Jacques Derrida, 'How to Avoid Speaking: Denials', in Harold Coward and Toby Foshay (eds) *Derrida and Negative Theology* (Albany NY: SUNY Press 1992) pp. 73–141.

31 See Jacques Derrida, '*Ousia* and *Gramme*: Note on a Note in *Being and Time*"', from Jacques Derrida, *Margins of Philosophy* [1972] trans. Alan Bass (Brighton, UK: Harvester Press, 1982).

32 For the development of Heidegger's phenomenology in the 1920s, see especially Theodore Kisiel, *The Genesis of Heidegger's 'Being and Time'* (Berkeley and Los Angeles CA: University of California Press 1993).

33 For a collection of papers dedicated to exploring this possibility, see John Joughin and Simon Malpas (eds) *The New Aestheticism* (Manchester: Manchester University Press 2002).

34 For discussions of Heidegger, see Jacques Derrida, *Of Spirit: Heidegger and the Question* [1987] trans. Geoffrey Bennington and Rachel Bowlby (Chicago: University of Chicago Press 1989); James Phillips, *Heidegger's Volk: Between National Socialism and Poetry* (Stanford CA: Stanford University Press, 2005); Miguel de Beistegui, *Heidegger and the Political* (London: Routledge 1997); and Joanna Hodge, *Heidegger and Ethics* (London: Routledge 1995) esp. ch. 4.

5.2 Time in the name of the other

1 Emmanuel Levinas, *God, Death and Time: Lectures 1975–1976* [1993] trans. Bettina Bergo (Stanford CA: Stanford University Press 2000) cited as L: GDT, 1975. This section of my enquiries was added in response to an anonymous reader, who noted an inconclusiveness of the previous version, and identified the sketchy nature of the account given of the trajectory of *L'animal que donc je suis*. Alas, the latter lacuna has not been corrected. I should like to mark a debt to this reader's perspicacity.

2 See Jacques Derrida, 'Typewriter Ribbon: Limited Ink (2)' [2001] trans. Peggy Kamuf, in Jacques Derrida, *Without Alibi* (Stanford CA: Stanford University Press 2002) pp. 71–160. An initial version of this appeared in Tom Cohen (ed.) *Paul de Man: Material Events and the After Life of Theory* (Minneapolis MN: University of Minnesota Press 2000). The former is a translation of the revised version, in Jacques Derrida, *Papier Machine* (Paris: Editions Galilée 2001).

3 See Jacques Derrida, 'Tense', trans. David Farrell Krell, in Kenneth Maly (ed.) *The Path of Archaic Thinking: Unfolding the Work of John Sallis* (Evanston IL: Northwestern University Press 1995) pp. 49–74.

4 Hannah Arendt attends to the open-ended temporality of forgiveness and to its strict impossibility in *The Human Condition* (Chicago: University of Chicago Press 1958).

5 Philippe Lacoue-Labarthe, 'The *Caesura* of the Speculative', in *Typography: Mimesis, Philosophy, Politics*, trans. and ed. Christopher Fynsk (Cambridge MA: Harvard University Press 1989) with 'Introduction: Desistance', pp. 1–42, by Jacques Derrida.

6 Jacques Derrida, *Apprendre à vivre enfin: entretien avec Jean Birnbaum* (Paris: Editions Galilée/Le Monde 2005) p. 54.

7 For a state of the art discussion of Kant on auto-poiesis and self-organisation, see Alberto Toscano, *The Theatre of Production: Philosophy and Individuation between Kant and*

Deleuze (London: Palgrave Macmillan 2006) especially part I: 'Kant beyond Kant or, the anomalies of the organic'.

8 Edmund Husserl, *On the Phenomenology of the Consciousness of Internal Time (1893–1917)* cited as HUE 4, trans. John Barnet Brough (Dordrecht: Kluwer Academic Publishers 1991).

9 Paul de Man, 'The Rhetoric of Temporality', in de Man, *Blindness and Insight: Essays in the Rhetoric of Contemporary Criticism* (Minneapolis MN: University of Minnesota Press 1971, 1983) pp. 187–228.

10 For Levinas' take on these paradoxical intentionalities, see the essays 'Intentionality and Metaphysics' (1959) and 'Intentionality and Sensation' (1965) in Emmanuel Levinas, *Discovering Existence with Husserl*, trans. Richard A. Cohen and Michael B. Smith (Evanston IL: Northwestern University Press 1998).

11 Jacques Derrida, *Rogue States: Two Essays on Reason* [2003] trans. Pascale Anne Brault and Michael Naas (Stanford CA: Stanford University Press 2005).

12 For a discussion of this relation, see Joanna Hodge, 'Derrida's Transcendental Contraband', *Textual Practice*, special edition on Derrida, ed. Simon Morgan Wortham, vol. 21, no. 2, May 2007.

13 See Marie-Louise Mallet (ed.) *L'animal autobiographique: autour de Jacques Derrida* (Paris: Editions Galilée 1999) in which the first part of the ten-hour lecture cycle appeared. See also Jacques Derrida, *L'animal que donc je suis*, texte établie par Marie-Louise Mallet (Paris: Editions Galilée 2006) which contains the more complete text.

14 This notion of legitimating fictions, borrowed from Michel de Montaigne, is explicitly averred to in Jacques Derrida, 'The Force of Law: On the Mystical Foundation of Authority', in Jacques Derrida, *Acts of Religion*, ed. Gil Anidjar (London: Routledge 2002) pp. 228–98.

15 For a discussion of this, see the interview between Jacques Derrida and Alexander Garcia Duttman, 'Perhaps or Maybe', in Jonathan Dronsfield and Nick Midgley (eds) *Responsibilities of Deconstruction, PLI Warwick Journal of Philosophy*, vol. 6, summer 1997, pp. 1–18.

16 See Paul Ricoeur, *Time and Narrative*, 3 vols [1983] trans. Kathleen McLaughlin and David Pellauer (Chicago: University of Chicago Press 1984–).

17 For an account of how the time of Aristotle's *Physics* may not be the time of his *De Motu Animalium*, see Martha Craven Nussbaum, *Aristotle's De Motu Animalium* (Princeton NJ: Princeton University Press 1978).

18 See Sigmund Freud, Penguin Freud Library, vol. 11, *On Meta-psychology and the Theory of Psychoanalysis*, ed. Angela Richards, trans. James Strachey (Harmondsworth, UK: Penguin 1984) 'The Unconscious', pp. 159–222.

19 See Jacques Derrida, *Archive Fever: A Freudian Impression* [1995] trans. Eric Prenowitz (Chicago: University of Chicago Press 1996). For a discussion of this text and of the manner in which it interrupts Derrida's writings on Husserl, some thirty years after their composition, see Joanna Hodge, 'Husserl, Derrida, *à suivre*', *Journal of the British Society for Phenomenology*, vol. 36, no. 2, May 2005.

20 See Emmanuel Levinas, *Difficile Liberté* [1963] 3rd edn, revised and corrected (Paris: Albin Michel Livre de Poche 1976) pp. 231–35, as discussed by Derrida, in Derrida, *L'animal que donc je suis*, in the context of a brief digression on Levinas' use of exclamation marks, pp. 157–60.

21 For Bergson, see Henri Bergson, *Time and Free Will: An Essay on the Immediate Data of Consciousness* [1889] trans. F. L. Pogson (London: George Allen and Unwin 1910); and *Duration and Simultaneity: Bergson and the Einsteinian Universe* [2nd edn 1923] trans. Leon Jacobsen, ed. Robin Durie (Manchester: Clinamen Press 1999). For Gaston Bachelard, see *The Dialectic of Duration* [1950] trans. Mary McAllester Jones (Manchester: Clinamen Press 2000). See also Keith Ansell Pearson, *Philosophy and the Adventure of the Virtual: Bergson and the Time of Life* (London: Routledge 2002); and Leonard Lawlor, *The Challenge of Bergsonism* (London: Continuum 2003). See also Robin Durie (ed.) *Time and the Instant* (Manchester: Clinamen Press 1999).

22 These reservations are articulated by Levinas in the opening sections of his *Existence and Existents* [1948] trans. Alphonso Lingis (The Hague: Martinus Nijhoff 1978).

23 See Jacques Derrida, *Adieu: To Emmanuel Levinas* [1997] trans. Pascale Anne Brault and Michael Naas (Stanford CA: Stanford University Press 1999).

24 See Catherine Malabou and Jacques Derrida, *Counterpath: Traveling with Jacques Derrida* [1999] trans. David Wills (Stanford CA: Stanford University Press 2004).

25 See Jacques Derrida, 'The Time of the Thesis: Punctuations', in Jacques Derrida, *Eyes of the University: Right to Philosophy 2* [1990] trans. Jan Plug *et al.* (Stanford CA: Stanford University Press 2004) pp.113–27. This paper was first published in English in Alan Montefiore (ed.) *Philosophy in France Today* (Cambridge: Cambridge University Press 1983) pp. 34–50.

Bibliography

Derrida: texts in French

Parages (Paris: Editions Galilée 1986).

Psyche (Paris: Editions Galilée, 1st edn 1987; 2nd edn: *Psyche: inventions de l'autre*, vol. 1 (1998) and vol. 2 (2001)).

Beliers: le dialogue ininterrompu entre deux infinis, le poème (Paris: Editions Galilée 2003).

Penser à Strasbourg, Derrida with Jean-Luc Nancy, Lucien Braun and Philippe Lacoue-Labarthe (Paris: Editions Galilée 2004).

Apprendre a vivre enfin: entretien avec Jean Birnbaum (Paris: Editions Galilée/Le Monde 2005).

'Jacques Derrida', *Cahier de l'Herne*, no. 83, eds Marie Louise Mallet and Ginette Michaud (Paris: Editions de l'Herne 2005).

Salut à Jacques Derrida, ed. Bruno Clement, *Rue Descartes*, 48 (2005), Collège Internationale de Philosophie.

L'animal que don je suis, texte établie par Marie Louise Mallet (Paris: Editions Galilée 2006).

Derrida: texts in English

Speech and Phenomena and Other Essays on Husserl's Theory of Signs [1967] trans. and ed. David Allison (Evanston IL: Northwestern University Press 1973).

Of Grammatology [1967] trans. Gayatri Chakravorty Spivak (Baltimore MD: Johns Hopkins University Press 1975).

Introduction to Husserl's 'Origin of Geometry' [1962] trans. John P. Leavey Jr (Brighton, UK: Harvester 1978).

Truth in Painting [1978], trans. Ian McLeod and Geoffrey Bennington (Chicago: University of Chicago Press 1987).

Writing and Difference [1967] trans. Alan Bass (London: Routledge 1978).

Positions, trans. and annotated by Alan Bass (Chicago: University of Chicago Press 1981).

Margins of Philosophy [1972] trans. Alan Bass (Brighton, UK: Harvester 1982).

The Ear of the Other: Otobiography Transference Translation, ed. Christie Macdonald (New York: Schocken Books 1985).

Glas: What Remains of Absolute Knowledge [1974] trans. John P. Leavey Jr (Lincoln NE: University of Nebraska Press 1986).

Memoires for Paul de Man [1984, 1986] trans. Jonathan Culler, Cecile Lindsay and Eduardo Cadava (London and New York: Columbia University Press 1986, 1989).

The Post Card: From Socrates to Freud and Beyond [1980] trans. Alan Bass (Chicago: University of Chicago Press 1987).

Limited Inc., trans. Sam Weber and Jeffrey Mehlman (Evanston IL: Northwestern University Press 1988).

A Derrida Reader: Between the Blinds, ed. Peggy Kamuf (London and New York: Harvester Wheatsheaf 1991).

Of Spirit: Heidegger and the Question [1987] trans. Geoffrey Bennington and Rachel Bowlby (Chicago: University of Chicago Press 1991).

Acts of Literature, ed. Derek Attridge (London and New York: Routledge 1992).

Given Time 1: Counterfeit Money [1991] trans. Peggy Kamuf (Chicago: University of Chicago Press 1992).

The Other Heading: Reflections on Today's Europe [1991] trans. Pascale Anne Brault and Michael Naas (Chicago: University of Chicago Press 1992).

Aporias: Dying – Awaiting (One Another at) the 'Limits of Truth' [1993] trans. Thomas Dutoit (Stanford CA: Stanford University Press 1993).

Specters of Marx: The State of the Debt, The Work of Mourning and the New International [1993] trans. Peggy Kamuf (Chicago: University of Chicago Press 1994).

On the Name, ed. Thomas Dutoit (Stanford CA: Stanford University Press 1995).

Points: Interviews 1974–1994, ed. Elisabeth Weber, trans. Peggy Kamuf *et al.* (Stanford CA: Stanford University Press 1995).

The Gift of Death [1992] trans. David Wills (Chicago: University of Chicago Press 1995).

Archive Fever: A Freudian Impression [1995] trans. Eric Prenowitz (Chicago: University of Chicago Press 1996).

Resistances of Psychoanalysis [1996] trans. Peggy Kamuf, Pascale Anne Brault and Michael Naas (Stanford CA: Stanford University Press 1996).

Politics of Friendship [1994] trans. George Collins (London: Verso 1997).

Adieu: To Emmanuel Levinas [1996] trans. Pascale Ann Brault and Michael Naas (Stanford CA: Stanford University Press 1999).

Monolingualism of the Other: Of the Original Prosthesis [1997] trans. Patrick Mensah (Stanford CA: Stanford University Press 1999).

The Problem of Genesis in Husserl's Philosophy [1990] trans. Marion Hobson (Chicago: University of Chicago Press 2001).

Who's Afraid of Philosophy: Right to Philosophy Volume One [1990] trans. Jan Plug (Stanford CA: Stanford University Press 2001–).

Without Alibi [2000] ed. and trans. Peggy Kamuf (Stanford CA: Stanford University Press 2001).

Acts of Religion, ed. Gil Anidjar (London and New York: Routledge 2002).

Negotiations: Interventions and Interviews 1971–2001, edited, translated and with an introduction by Elizabeth Rottenberg (Stanford CA: Stanford University Press 2002).

The Work of Mourning, trans. and eds Pascale Anne Brault and Michael Naas (Chicago: University of Chicago Press 2002).

Eyes of the University: Right to Philosophy 2 [1990] trans. Jan Plug *et al.* (Stanford CA: Stanford University Press 2004).

On Touching: Jean-Luc Nancy, trans. Christine Irizarry (Stanford CA: Stanford University Press 2005).

Paper Machine [2001] trans. Rachel Bowlby (Stanford CA: Stanford University Press 2005).

Rogue States: Two Essays on Reason [2003] trans. Pascale Anne Brault and Michael Naas (Stanford CA: Stanford University Press 2005).

Joint authorship

Bennington, Geoffrey and Derrida, Jacques: *Jacques Derrida* (1991) trans. Geoffrey Bennington (Chicago: University of Chicago Press).

Blanchot, Maurice, and Derrida, Jacques: *The Instant of My Death* [1994]; *Demeure* [1998] trans. Elizabeth Rottenberg (Stanford CA: Stanford University Press 2000).

Derrida, Jacques and Ferraris, Maurizio: *A Taste for the Secret* [1997] trans. Giacomo Donis (Cambridge: Polity Press 2001).

Derrida, Jacques and Stiegler, Bernard: *Echographies of Television* (Cambridge: Polity Press 2002).

Dufourmantelle, Anne and Derrida, Jacques: *Of Hospitality* [1997] trans. Rachel Bowlby (Stanford CA: Stanford University Press 2000).

Malabou, Catherine and Derrida, Jacques: *Counterpaths* [1997] trans. David Wills (Stanford CA: Stanford University Press 2004).

Roudinesco, Elisabeth and Derrida, Jacques: *For What Tomorrow? A Dialogue*, trans. Jeff Fort (Stanford CA: Stanford University Press 2004).

Other texts consulted

Agamben, Giorgio: *Homo Sacer: Sovereign Power and Bare Life* [1994] trans. Daniel Heller Roazen (Stanford CA: Stanford University Press 1998).

Ansell-Pearson, Keith: *Germinal Life: The Difference an Repetition of Deleuze* (London: Routledge 1999).

—— *Philosophy and the Adventure of the Virtual: Bergson and the Time of Life* (London: Routledge 2002).

Aristotle: *De Anima*, trans. Hugh Lawson Tancred (Harmondsworth, UK: Penguin 1986).

—— *Physics*, trans. Robin Waterfield, introduction and notes by David Bostock, Oxford World's Classics (Oxford: Oxford University Press 1996); also see Edward Hussey (ed.): *Physics Books III and IV*, Clarendon Aristotle series, general editor J. L. Ackrill (Oxford: Clarendon Press 1983).

—— *Metaphysics*, trans. Hugh Lawson Tancred (Harmondsworth, UK: Penguin 1998).

Augustine: *Confessions*, trans. Henry Chadwick, Oxford World's Classics (Oxford: Oxford University Press 1991).

Bachelard, Gaston: *The Dialectic of Duration* [1950] trans. Mary McAllester Jones (Manchester: Clinamen 2000).

Baker, Peter: *Deconstruction and the Ethical Turn* (Gainesville FL: University of Florida Press 1995).

Banham, Gary (ed.): *Husserl and the Logic of Experience* (London: Macmillan Palgrave 2005).

—— *Husserl and Derrida*, special number of the *Journal of the British Society for Phenomenology*, vol. 26, no. 2, May 2005.

Banham, Gary and Blake, Charlie (eds): *Evil Spirits: Nihilism and the Fate of Modernity* (Manchester: Manchester University Press 2000).

Beardsworth, Richard: *Derrida and the Political* (London: Routledge 1996).

Benjamin, Andrew (ed.): *Walter Benjamin and Art* (London: Continuum 2005).

Benjamin, Andrew and Hanssen, Beatrice (eds): *Walter Benjamin and Romanticism* (London: Continuum 2002).

Benjamin, Andrew and Osborne, Peter (eds): *Walter Benjamin's Philosophy: Destruction and Experience*, 2nd edn (Manchester: Clinamen Press 2000).

Benjamin, Walter: *Selected Writings*, 4 vols, eds Marcus Bullock and Michael W. Jennings (Cambridge MA: The Belknap Press of Harvard University 1996–).

Bergson, Henri: *Time and Free Will: An Essay on the Immediate Data of Consciousness* [1910] trans. F. L.Pogson (London: George Allen and Unwin 1989).

—— *Duration and Simultaneity* [1923] trans. Leon Jacobsen (Manchester: Clinamen Press 1999).

Bernasconi, Robert and Critchley, Simon (eds): *Re-reading Levinas* (London: Athlone Press 1991).

Bernet, Rudolf: 'Perception as a Teleological Process of Cognition', in *Analecta Husserliana vol. IX*, ed. Anna Teresa Tymieniecka (Dordrecht: Reidel 1979).

—— 'Husserl's Theory of Signs Revisited', in Robert Sokolowski (ed.) *Edmund Husserl and the Phenomenological Tradition: Studies in Philosophy and the History of Philosophy no. 18* (Washington DC: Catholic University of America Press 1988).

—— 'Derrida-Husserl-Freud: The Trace of Transference', in Spindel Conference 1993: Derrida's Interpretation of Husserl, supplement to the *Southern Journal of Philosophy*, no. 32, 1994.

—— 'An Intentionality without Subject of Object?', *Man and World*, vol. 27, 1994, pp. 231–55.

Bernet, Rudolf, Marbach, Eduard and Kern, Iso: *Introduction to Husserlian Phenomenology* (Evanston IL: Northwestern University Press 1995).

Bident, Christophe and Vilar, Pierre (eds): *Maurice Blanchot: récits critiques* (Paris: Editions Farrago/Editions Leo Scheer 2003).

Blanchot, Maurice: *The Siren's Song*, ed. Gabriel Josipovici (Brighton, UK: Harvester Press 1982).

—— *The Space of Literature*, trans. Ann Smock (Lincoln NE and London: University of Nebraska Press 1982).

—— *Vicious Circles: Two Fictions & 'after the fact'*, trans. Paul Auster (Barrytown NY: Station Hill Press 1985).

—— *The Writing of the Disaster* [1980] trans. Ann Smock (Lincoln NE and London: University of Nebraska Press 1986).

—— *The Step (not) Beyond* [1972] trans. Lycette Nelson (Albany NY: SUNY Press 1992).

—— *The Infinite Conversation* [1969] trans. Susan Hanson (Minneapolis MN: University of Minnesota Press 1993).

—— *The Work of Fire* [1949] trans. Charlotte Mandell (Stanford CA: Stanford University Press 1995).

—— *Friendship* [1971] trans. Elizabeth Rottenberg (Stanford CA: Stanford University Press 1997).

Blattner, Wilhelm: *Heidegger's Temporal Idealism* (Cambridge: Cambridge University Press 1999).

Bloom, Harold, de Man, Paul, Derrida, Jacques, Hartman, Geoffrey and Hillis Miller, J.: *Deconstruction and Criticism* (New York: Continuum 1979).

Borradori, Giovanna (ed.): *Philosophy in a Time of Terror: Dialogues with Juergen Habermas and Jacques Derrida* (Chicago: University of Chicago Press 2003).

Bradley, Arthur: *Negative Theology and Modern French Philosophy* (London: Routledge 2004).

Brown, Peter: *Augustine of Hippo: A Biography* (London: Faber 1967).

Bruns, Gerald L.: *Heidegger's Estrangements: Language, Truth and Poetry in the Later Writings* (New Haven CT: Yale University Press 1989).

—— *Maurice Blanchot and the Refusal of Philosophy* (Baltimore MD: Johns Hopkins University Press 1997).

Bruzina, Rudolf: *Edmund Husserl and Eugen Fink: Beginnings and Ends in Phenomenology 1928–1938* (New Haven CT: Yale University Press 2004).

Byers, Damian: *Intentionality and Transcendence: Closure and Openness in Husserl's Phenomenology* (Seattle WA: Noesis Press, University of Wisconsin Press 2002).

Caputo, John and Scanlon, Michael (eds): *God the Gift and Postmodernism* (Bloomington IN: Indiana University Press 1999).

Carr, David: *Phenomenology and the Problem of Hstory* (Dordrecht: Kluwer 1974).

Caygill, Howard: *Walter Benjamin: The Colour of Experience* (London: Routledge 1998).

Cohen, Margaret: *Profane Illumination: Walter Benjamin and the Paris of Surrealist Revolution* (Berkeley CA: University of California Press 1993).

Cohen, Richard (ed.): *Face to Face with Levinas* (Albany NY: SUNY Press 1986).

Cohen, Tom (ed.): *Jacques Derrida and the Humanities: A Critical Reader* (Cambridge: Cambridge University Press 2001).

—— *Material Events: Paul de Man and the Afterlife of Theory* (Minneapolis MN: University of Minnesota Press 2001).

Coward, Harold and Foshay, Toby (eds): *Derrida and Negative Theology* (Albany NY: SUNY Press 1992).

Cixous, Hélène: *Portrait of Jacques Derrida as a Young Jewish Saint* [2001] trans. Beverley Bie Brihac (New York: Columbia University Press 2004).

Critchley, Simon: *The Ethics of Deconstruction: Derrida and Levinas* (Oxford: Blackwell 1992).

Critchley, Simon, and Bernasconi, Robert: *The Cambridge Companion to Levinas* (Cambridge: Cambridge University Press 1998).

Crowell, Stephen: *Husserl, Heidegger and the Space of Meaning* (Evanston IL: Northwestern University Press 2001).

Dastur, Françoise: *Death and Essay on Finitude* [1989] Athlone Contemporary European Philosophy series (London and New Brunswick NJ: Athlone 1996).

—— *Telling Time: Sketch of a Phenomenological Chrono-logy* [1994] trans. Edward Bullard, Athlone Contemporary European Thinkers series (London and New Brunswick NJ: Athlone Press 2000).

De Beistegui, Miguel: *Heidegger and the Political* (London: Routledge 1999).

—— *The New Heidegger* (London: Continuum 2005).

De Boer, Karin: *Thinking in the Light of Time: Heidegger's Encounter with Hegel* [1997] trans. Karin de Boer and Janet Taylor (Albany NY: SUNY Press 2000).

De Man, Paul: *Blindness and Insight: Essays in the Rhetoric of Contemporary Criticism* (Minneapolis MN: University of Minnesota Press 1971, 2nd edn 1983).

—— *Allegories of Reading: Figural Language in Rousseau, Nietzsche, Rilke and Proust* (New Haven CT: Yale University Press 1979).

—— *The Resistance to Theory* (Minneapolis MN: University of Minnesota Press 1986).

—— *Aesthetic Ideology*, ed. Andrej Warminski (Minneapolis MN: University of Minnesota Press 1996).

De Vries, Hent: *Philosophy and the Turn to Religion* (Baltimore MD: Johns Hopkins University Press 1999).

—— *Religion and Violence: Philosophical Perspectives from Kant to Derrida* (Baltimore MD: Johns Hopkins University Press 2002).

Dronsfield, Jonathan and Midgley, Nick (eds): *Responsibilities of Deconstruction, PLI Warwick Journal of Philosophy*, vol. 6, summer 1997.

Duettman, Alexander Garcia: *The Gift of Language: Memory and Promise in Adorno, Benjamin, Heidegger and Rosenzweig* [1989] Athlone Contemporary European Thinkers series (London and New Brunswick NJ: Athlone Press 2000).

Fenves, Peter: *A Peculiar Fate: Metaphysics an World History in Kant* (Ithaca NY: Cornell University Press 1991).

Fenves, Peter (ed.): *Raising the Tone of Philosophy: Late Essays by Immanuel Kant; Transformative Critique by Jacques Derrida* (Baltimore MD: Johns Hopkins University Press 1993).

Ferris, David S. (ed.): *Walter Benjamin: Theoretical Questions* (Stanford CA: Stanford University Press 1994).

—— *The Cambridge Companion to Walter Benjamin* (Cambridge: Cambridge University Press 2004).

Fink, Eugen: *Sixth Cartesian Meditations: The Idea of a Transcendental Theory of Method*, with textual annotations by Edmund Husserl [1988] trans. Ronald Bruzina, (Bloomington IN: Indiana University Press 1995).

Fioretis, Aris: *Word Traces: Readings of Paul Celan* (Baltimore MD: Johns Hopkins University Press 1994).

Freud, Sigmund: *Case Histories II*, trans. James Strachey, ed. Angela Richards, Penguin Freud Library vol. 9 (Harmondsworth, UK: Penguin 1979).

—— *On Metapsychology: The Theory of Psychoanalysis*, trans. James Strachey, ed. Angela Richards, Penguin Freud Library vol. 11 (Harmondsworth, UK: Penguin 1984).

Fukuyama, Francis: *The End of History and the Last Man* (London: Macmillan 1992).

Gallagher, Shaun: *The Inordinance of Time* (Evanston IL: Northwestern University Press 1998).

Gasche, Rodolphe: *The Tain of the Mirror: Derrida and the Philosophy of Reflection* (Cambridge MA: Harvard University Press 1986).

—— *The Wild Card of Reading: On Paul de Man* (Cambridge MA: Harvard University Press 1998).

—— *Inventions of Difference: On Jacques Derrida* (Cambridge MA: Harvard University Press 1994).

Gaston, Sean: *Derrida and Disinterest* (London: Continuum 2005).

Gershom Scholem, A. and Adorno, Theodor W. (eds): *Walter Benjamin Briefe 1 und 2* (Frankfurt am Main, Germany: Suhrkamp Verlag 1978).

Gibbs, Robert: *Correlations in Rosenzweig and Levinas* (Princeton NJ: Princeton University Press 1992).

Hamacher, Werner: *Premises – Reading in Hegel: The Genesis and Structure of a Dialectical Hermeneutics* [1978] trans. Nicholas Walker and Simon Jarvis (London: Athlone Press 1998).

—— '"Lingua Amissa: The Messianism of Commodity Language" and Derrida's "Specters of Marx"', in Richard Rand (ed.) *Futures: Of Jacques Derrida* (Stanford CA: Stanford University Press 2001), pp. 130–77.

—— '"Now": Walter Benjamin and Historical Time', in Heidrun Friese (ed.) *The Moment: Time and Rupture in Modern Thought* (Liverpool: Liverpool University Press 2001).

—— 'Cult History: Benjamin's Sketch "Capitalism as Religion"', *Diacritics: Journal of Contemporary Criticism*, vol. 3, nos. 3–4 (2002) pp. 81–106.

Hand, Sean (ed.): *The Levinas Reader* (Oxford: Blackwell 1993).

Hanssen, Beatrice: *Walter Benjamin's Other History: Of Stones, Animals, Human Being and Angels* (Berkeley CA: University of California Press 1998).

—— *Critique of Violence: Between Post-structuralism and Critical Theory* (London: Routledge 2000).

Hanssen, Beatrice and Benjamin, Andrew (eds): *Walter Benjamin and Romanticism* (London: Continuum 2002).

Hegel, G. W. F.: *The Philosophy of Mind*, part III of the *Encyclopaedia of the Philosophical Sciences* [1830] trans. William Wallace (Oxford: Clarendon Press 1971).

Heidegger, Martin: *Being and Time* [1927] trans. John Macquarrie and Edward Robinson (Oxford: Blackwell 1962).

—— *Metaphysical Foundations of Logic Beginning with Leibniz* [SS 1928, GA 26] trans. Michael Heim (Bloomington IN: Indiana University Press 1984).

—— *Basic Problems of Phenomenology* [1927, GA 24] trans. Alfred Hofstadter (Bloomington IN: Indiana University Press 1988).

—— *The Fundamental Concepts of Phenomenology: World, Finitude, Solitude* [1929–30, GA 29–30] trans. Will McNeill and Nicholas Walker (Bloomington IN: Indiana University Press 1995).

—— *History of the Concept of Time* [1925, GA 20] trans. Theodore Kisiel (Bloomington IN: Indiana University Press 1997).

—— *The Concept of Time* [1924] trans. Will McNeill (Cambridge: Cambridge University Press 1998).

—— *Pathmarks* [1976, GA 9] trans. Willliam H. McNeill (Cambridge: Cambridge University Press 1998).

—— *Contributions to Philosophy: Of Enowning*, trans. Parvis Emad and Kenneth Maly (Bloomington IN: Indiana University Press 2001).

—— *Phenomenological Interpretation of Aristotle: Initiation into Phenomenological Research* [1921–22, GA 61] trans. Richard Rojcewicz (Bloomington IN: Indiana University Press 2001).

—— *Off the Beaten Track* [1949] trans. Kenneth Haynes and Julian Young (Cambridge: Cambridge University Press 2002).

—— *Phenomenology of Religious Life* [1922, GA 62] trans. Matthias Fritsch and Jennifer Anna Gosetti-Ferencei (Bloomington IN: Indiana University Press 2002).

—— *Introduction to Phenomenological Research* [1923–24, GA 17] trans. Daniel Dahlstrom, (Bloomington IN: Indiana University Press 2005).

Hobson, Marion: *Opening Lines: Jacques Derrida* (London and New York: Routledge 1997).

Hodge, Joanna: *Heidegger and Ethics* (London: Routledge 1995).

—— 'Ethics and Time: Levinas between Kant and Husserl', *Diacritics: Journal for Contemporary Criticism*, vol. 32, no. 3, 2002, pp. 107–35.

—— 'Why Aesthetics Might Be Several: On Nancy and Technics', in special issue of *Angelaki: Journal of the Theoretical Humanities*, ed. Gary Banham, 'On the Ends of Aesthetics', vol. 7, no. 1, 2002.

—— 'Husserl, Derrida, *à Suivre*' in special issue on Husserl and Derrida, *Journal of the British Society for Phenomenology*, vol. 36, no. 2, May 2005.

—— 'The Timing of Elective Affinity: Walter Benjamin's Strong Aesthetics', in Andrew Benjamin (ed.) *Walter Benjamin and Art* (London: Continuum 2005) pp. 14–31.

—— 'Derrida's Transcendental Contraband', in special number on Derrida, *Textual Practice*, vol. 21, no. 2, May 2007.

——'Phenomenology to come: Derrida's ellipses' in Martin McQuillan (ed.) *The Politics of Deconstruction: Jacques Derrida and the Other of Philosophy* (London: Pluto Press 2007).

Hoeffe, Otfried: *Kant's Cosmopolitan Theory of Law and Peace* [2001] trans. Alexandra Newton (Cambridge: Cambridge University Press 2006).

Howells, Christina: *Derrida* (Cambridge: Polity 1998).

Hill, Leslie: *Blanchot; Extreme Contemporary* (London: Routledge 1998).

Husserl, Edmund: *Formal and Transcendental Logic* [1928, HUA 17] trans. Dorion Cairns (The Hague: Martinus Nijhoff 1969).

—— *The Crisis of the European Sciences and Transcendental Philosophy* [1938, HUA 6] trans. David Carr (Evanston IL: Northwestern University Press 1970).

—— *Logical Investigations* [1900–01, 1913–21, HUA 18, 19, 20] trans. J. F. Findlay (London: Routledge 1970).

—— *Cartesian Meditations: An Introduction to Phenomenology* [1931, HUA 1] trans. Dorion Cairns (The Hague: Martinus Nijhoff 1973).

—— *Ideas: A General Introduction to Pure Phenomenology* [1913, HUA 3, 1–2] trans. W. Boyce Gibson (New York: George Allen and Unwin 1931; London: Collier Macmillan 1974) commonly known as *Ideas One*.

—— *Ideas Pertaining to a Pure Phenomenology, Second Book: Studies in the Phenomenology of Constitution (Ideas Two)* [HUA 4] eds and trans. R. Rojcewicz and Andre Schuwer (Dordrecht: Kluwer Academic Publishers 1989).

—— *On the Phenomenology of the Consciousness of Internal Time (1893–1917)* [1928, HUA 10] trans. John Barnet Brough (Dordrecht: Kluwer Academic Publishers 1991).

—— *The Idea of Phenomenology* [1907, HUA 2] trans. Lee Hardy (Dordrecht: Kluwer Academic Publishers 1999).

—— *Analyses Concerning Passive and Active Synthesis: Lectures on Transcendental Logic* [1921-: HUA 11 and 31] trans. Antony Steinbock (Dordrecht: Kluwer 2004).

—— *Phantasy, Image Consciousness and Memory (1893–1933)* [HUA 23] trans. John Barnet Brough (Dordrecht: Springer Verlag 2005).

Husserl, Edmund and Landgrebe, Ludwig: *Experience and Judgment: Investigations in a Genealogy of Logic* [1938] trans. James Churchill and Karl Ameriks (Evanston IL: Northwestern University Press 1973).

Janicaud, Dominique: *Phenomenology and the 'Theological Turn': The French Debate* [1991] trans. Bernard G. Prusak (New York: Fordham University Press 2000).

Johnson, Christopher: *System and Writing in the Philosophy of Jacques Derrida* (Cambridge: Cambridge University Press 1993).

Joughin, John and Malpas, Simon (eds) *The New Aestheticism* (Manchester: Manchester University Press 2000).

Kamuf, Peggy: *The Division of Literature: Or the University in Deconstruction* (Chicago: University of Chicago Press 1997).

Kant, Immanuel: *The Critique of Pure Reason* [1781, 1787] trans. Norman Kemp Smith, (London: Macmillan 1929).

—— *Critique of Practical Reason* [1788] trans. Lewis White Beck (Indianapolis IN: Bobbs Merrill 1956).

—— *Anthropology from a Pragmatic Point of View* [1798] trans. Mary J. Gregor (The Hague: Martinus Nijhoff 1974).

—— *Groundwork for a Metaphysic of Morals* [1785] trans. James W. Ellington (Indianapolis IN: Hackett 1981).

—— *Critique of Judgment* [1790] trans. Werner S. Pluhar (Indianapolis IN: Hackett 1987).

—— *Metaphysics of Morals* [1797] trans. and ed. Mary J. Gregor, Cambridge Texts in the History of Philosophy (Cambridge: Cambridge University Press 1996).

—— *Religion within the Boundaries of Mere Reason and Other Writings*, eds Allen Wood and George di Giovanni, *Cambridge Texts in the History of Philosophy* (Cambridge: Cambridge University Press 1998).

Kates, Joshua: *Essential History: Jacques Derrida and the Development of Deconstruction* (Evanston IL: Northwestern University Press 2005).

Kisiel, Theodore: *The Genesis of Heidegger's 'Being and Time'* (Berkeley CA: University of California Press 1993).

Kortooms, Toine: *Phenomenology of Time: Edmund Husserl's Analysis of Time-Consciousness* (Dordrecht: Kluwer Academic Publishers 2002).

Kronick, Joseph G.: *Derrida and the Future of Literature* (Albany NY: SUNY Press 1999).

Lacan, Jacques: *Ecrits* [1966] trans. Alan Sheridan (London: Tavistock Press 1977).

Lacoue-Labarthe, Philippe: *Typography: Mimesis Philosophy Politics*, ed. and trans. Christopher Fynsk (Cambridge MA: Harvard University Press 1989).

Lacoue-Labarthe, Philippe and Nancy, Jean-Luc: *The Literary Absolute: Theory of Literature in German Romanticism* [1972] trans. Philip Barnard and Cheryl Lester (Albany NY: SUNY Press 1988).

Large, Will: *Emmanuel Levinas and Maurice Blanchot: Ethics and the Ambiguity of Writing* (Manchester: Clinamen 2005).

Lawlor, Leonard: *Derrida and Husserl: The Basic Problem of Phenomenology* (Bloomington IN: Indiana University Press 2002).

—— *The Challenge of Bergsonism* (London: Continuum 2003).

Leask, Ian and Cassidy, Eoin: *Givenness and God: Questions of Jean Luc Marion* (New York: Fordham University Press 2005).

Leibniz, Georg: *Discourse on Metaphysics and Other Essays*, trans. Daniel Garber and Roger Ariew (Indianapolis IN: Hackett 1991).

Levinas, Emmanuel: *Totality and Infinity: An Essay on Exteriority* [1961] trans. Alphonso Lingis (Pittsburgh PA: Duquesne University Press 1969).

—— *Difficile liberté* [1963] 3rd edn (Paris: Presse Albin Michel 1976).

—— *Otherwise than Being, or, Beyond Essence* [1974] trans. Alphonso Lingis (Pittsburgh PA: Duquesne University Press 1978).

—— *Discovering Existence with Husserl*, trans. Richard Cohen and Michael B. Smith (Evanston IL: Northwestern University Press 1995).

—— *Alterity and Transcendence* [1995] trans. Michael B. Smith (London: Athlone Press 1999).

—— *Existence and Existents* [1948] trans. R. Bernasconi and Alphonso Lingis (Pittsburgh PA: Duquesne University Press 2001)

—— *Of God Who Comes to Mind* [1996] trans. Bettina Bergo (Stanford CA: Stanford University Press 2001).

—— *God, Death and Time: Lectures 1975–1976* [1993] trans. Bettina Bergo (Stanford CA: Stanford University Press 2002).

—— *Entre Nous: Thinking-of-the-Other*, trans. Michael B. Smith (London: Continuum 2003).

Librett, Jeffrey (ed.): *Of the Sublime: Presence in Question* [1988] (Albany NY: SUNY Press 1993).

Llewelyn, John: *Derrida and the Threshold of Sense* (London: Macmillan 1986).

—— *The Middle Voice of Ecological Conscience: A Chiasmic Reading of Responsibility in the Neighbourhood of Levinas, Heidegger and Others* (New York: St Martin's Press 1991).

—— *Appositions of Jacques Derrida and Emmanuel Levinas* (Bloomington IN: Indiana University Press 2002).

Lyotard, J-F.: *The Inhuman: Reflections on Time* [1988] trans. Geoffrey Bennington and Rachel Bowlby (Oxford: Polity Press 1991).

—— *Heidegger and the 'Jews'* [1988] trans. Andreas Michel (Minneapolis MN: University of Minnesota Press 1993).

—— *The Differend: Phrases in Dispute* [1983] trans. George van den Abeele (Minneapolis MN: University of Minnesota Press 1993).

—— *Lessons on the Analytic of the Sublime* [1991] trans. Elizabeth Rottenberg (Stanford CA: Stanford University Press 1994).

—— *The Confession of Augustine* [1998] trans. Richard Beardsworth (Stanford CA: Stanford University Press 2000).

Malabou, Catherine: *The Future of Hegel: Plasticity, Temporality and the Dialectic* [1996] trans. Lisabeth During (London: Routledge 2004).

Maly, Kenneth (ed.): *The Path of Archaic Thinking: Unfolding the Work of John Sallis* (Evanston IL: Northwestern University Press 1995).

Marion, J. L.: *God without Being* [1983] trans. Thomas A. Carlson (Chicago: University of Chicago Press 1991).

—— *Reduction and Givenness: Investigations of Husserl, Heidegger and Phenomenology* [1989] trans. Thomas A. Carlson (Evanston IL: Northwestern University Press 1998).

—— *Being Given: Toward a Phenomenology of Givenness* [1997] trans. Jeffrey L. Kosky (Stanford CA: Stanford University Press 2002).

Marrati, Paola: *Genesis and Trace: Derrida Reading Husserl and Heidegger* [1998] trans. Simon Sparks (Stanford CA: Stanford University Press 2005).

Marx, Karl and Engels, Friedrich: *The German Ideology* [1846] in Marx and Engels *Collected Works* (New York: International Publishers 1976).

McKenna, Bill and Evans, J. Claude (eds): *Derrida and Phenomenology* (Dordrecht: Kluwer Academic Publishers 1995).

Mensch, James: *Postfoundational Phenomenology: Husserlian Reflections on Presence and Embodiment* (University Park PA: Pennsylvania State University Press 2001).

Merleau-Ponty, Maurice: *Signs* [1960] trans. Richard McLeary (Evanston IL: Northwestern University Press 1964).

—— *The Phenomenology of Perception* [1945] trans. Colin Smith (London: Routledge 1978).

—— *Husserl at the Limits of Phenomenology: Including Texts by Edmund Husserl*, eds Len Lawlor and Bettina Bergo (Evanston IL: Northwestern University Press 2001).

Michelfelder, Diane P. and Palmer, Richard E.: *Dialogue and Deconstruction: The Gadamer–Derrida Encounter* (Albany NY: SUNY Press 1989).

Nancy, J-L.: *The Inoperative Community* [1986] trans. Peter Connor *et al.* (Minneapolis MN: University of Minnesota Press 1991).

—— *The Birth to Presence*, trans. Brian Holmes *et al.* (Stanford CA: Stanford University Press 1993).

—— *The Experience of Freedom* [1988] trans. Bridget McDonald (Stanford CA: Stanford University Press 1993).

—— *The Muses* [1994] trans. Peggy Kamuf (Stanford CA: Stanford University Press 1996).

—— *A Sense of the World* [1993] trans. Jeffrey S. Librett (Minneapolis MN: University of Minnesota Press 1997).

—— *Being Singular Plural* [1996] trans. Robert D. Richardson and Anne E. O'Byrne (Stanford CA: Stanford University Press 2002).

—— *A Finite Thinking* [1990] ed. Simon Sparks (Stanford CA: Stanford University Press 2003).

Nass, Michael: *Taking on Tradition: Jacques Derrida and the Legacies of Deconstruction* (Stanford CA: Stanford University Press 2003).

Neske, Gunther and Kettering, Emil (eds): *Martin Heidegger and National Socialism: Questions and Answers* [1988] trans. Lisa Harries (New York, NY: Paragon House 1990).

Nietzsche, Friedrich: *The Will to Power* [1901] ed. and trans. Walter Kaufmann (New York NY: Vintage Books 1968).

—— *On the Genealogy of Morals and Ecce Homo* [1886] trans. R. J. Hollingdale and Walter Kaufmann (New York NY: Vintage Books 1969).

Nussbaum, Martha Craven: *Aristotle's De Motu Animalium: Text and Commentary* (Princeton NJ: Princeton University Press 1978).

Phillips, James: *Heidegger's Volk: Between National Socialism and Poetry* (Stanford CA: Stanford University Press 2005).

Rand, Richard (ed.): *Futures of Jacques Derrida* (Stanford CA: Stanford University Press 2001).

Rapaport, Hermann: *Heidegger and Derrida: Reflections on Time and Language* (Lincoln NE and London: University of Nebraska Press 1989).

—— *Later Derrida* (London: Routledge 1999).

Ricoeur, Paul: *Husserl: An Analysis of His Phenomenology* (Evanston IL: Northwestern University Press 1967).

—— *The Rule of Metaphor: Multi-disciplinary Studies in the Creation of Meaning in Language* [1975] trans. Robert Czerny (London: Routledge and Kegan Paul 1978).

—— *Time and Narrative* [1983] 3 vols, trans. Kathleen McLoughlin and David Pellauer (Chicago: University of Chicago Press 1984–).

Rothfield, Philip (ed.): *Kant after Derrida* (Manchester: Clinamen 2004).

Royle, Nicholas: *Jacques Derrida*, Routledge Critical Thinkers series (London: Routledge 2003).

Royle, Nicholas (ed.): *Deconstruction: A Critical Reader* (London: Palgrave Macmillan 2000).

Sallis, John (ed.): *Deconstruction and Philosophy: The Texts of Jacques Derrida* (Chicago: University of Chicago Press 1987).

Sallis, John (ed.): *Reading Heidegger: Commemorations* (Bloomington IN: Indiana University Press 1993).

Smith, Joseph H. and Kerrigan, William (eds): *Taking Chances: Derrida, Psychoanalysis and Literature* (Baltimore MD: Johns Hopkins University Press 1984).

Smith, Robert: *Derrida and Autobiography* (Cambridge: Cambridge University Press 1994).

Sprinker, Michael (ed.): *Marx and Sons* (London: Verso 1997).

Steinbock, Antony: *Home and Beyond: Generative Phenomenology after Husserl* (Evanston IL: Northwestern University Press 1995).

Stiegler, Bernard: *Technics and Time: The Fault of Epimetheus* [1994] trans. Richard Beardsworth and George Collins (Stanford CA: Stanford University Press 1998).

Taminiaux, Jacques: *Dialectic and Difference: Modern Thought and the Sense of Human Limits*, trans. Robert Crease and James T. Decker (Atlantic Highlands NJ: Humanities Press 1985).

—— *Heidegger and the Project of Fundamental Ontology*, trans. Michael Gendre (Albany NY: SUNY Press 1991).

—— *Poetics, Speculation and Judgment: The Shadow of the Work of Art from Kant to Phenomenology*, trans. Michael Gendre (Albany NY: SUNY Press 1993).

Taylor, Mark (ed.): *Deconstruction in Context: Literature and Philosophy* (Chicago: University of Chicago Press 1982).

Toscano, Alberto: *The Theatre of Production: Philosophy and Individuation between Kant and Deleuze* (London: Palgrave Macmillan 2006).

Waters, Lindsay and Godzich, Wlad (eds): *Reading de Man Reading* (Minneapolis MN: University of Minnesota Press 1989).

Welton, Donn: *The Other Husserl* (Bloomington IN: Indiana University Press 2002).

Welton, Donn (ed.): *The New Husserl: A Critical Reader* (Indianapolis IN: Indiana University Press 2003).

Wood, David: *The Deconstruction of Time* (Atlantic Highlands NJ: Humanities Press 1989).

Wood, David (ed.): *Derrida: A Critical Reader* (Oxford: Blackwell 1993).

—— *Of Derrida, Heidegger and Spirit* (Evanston IL: Northwestern University Press 1993).

Wyschogrod, Edith: *Emmanuel Levinas: The Problem of Ethical Metaphysics* (New York NY: Fordham University Press 2000).

Zahavi, Dan: *Self-awareness and Alterity: A Phenomenological Investigation* (Evanston IL: Northwestern University Press 1999).

—— *Introduction to Phenomenology* (Stanford CA: Stanford University Press 2002).

—— 'Time and Consciousness in the Bernauer Manuscripts', *Husserl Studies*, vol. 20, no. 2 (2004) pp. 99–118.

Index